Studies in Computational Intelligence

Volume 564

Series editor

Janusz Kacprzyk, Polish Academy of Sciences, Warsaw, Poland
e-mail: kacprzyk@ibspan.waw.pl

Quan Bai • Fenghui Ren • Minjie Zhang
Takayuki Ito • Xijin Tang
Editors

Smart Modeling and Simulation for Complex Systems

Practice and Theory

Springer

Editors
Quan Bai
Auckland University of Technology
Auckland, New Zealand

Fenghui Ren
University of Wollongong
Wollongong, NSW, Australia

Minjie Zhang
School of Computer Science and Software
    Engineering
University of Wollongong
Wollongong, NSW, Australia

Takayuki Ito
Nagoya Institute of Technology
Nagoya, Japan

Xijin Tang
Academy of Mathematics and Systems
    Science
Chinese Academy of Science
Beijing, China

ISSN 1860-949X   ISSN 1860-9503 (electronic)
ISBN 978-4-431-56156-9   ISBN 978-4-431-55209-3 (eBook)
DOI 10.1007/978-4-431-55209-3
Springer Tokyo Heidelberg New York Dordrecht London

Preface

Computer based modelling and simulation have become useful tools to facilitate humans to understand systems in different domains, such as physics, chemistry, biology, economics, engineering and social science. A complex system is featured with a large number of interacting components, whose aggregate activities are nonlinear and self-organized. Complex systems are hard to be simulated or modelled by using traditional computational approaches due to the complex relationships of components and distributed features of resources, and dynamic work environments. Meanwhile, smart systems such as multi-agent systems have demonstrated advantages and great potentials in modelling and simulating complex systems.

This book constitutes extended versions of selected papers from the international workshop on Smart Simulation and Modelling for Complex Systems (SSMCS'13) in IJCAI'13, which was held in Beijing, China, in August 2013. Each paper was peer reviewed by three PC members. We solicited papers on all aspects of smart simulation and modelling of complex systems through using agent and other AI technologies. They are, for instance, being studied in the network modelling, microsimulation and modelling, social influence modelling, disaster modelling, environment modelling, power market modelling, and idea discovery process modelling.

Finally, we would like to extend our sincere thanks to all authors. This book would not have been possible without their valuable supports and contributions.

Auckland, New Zealand
Wollongong, NSW, Australia
Wollongong, NSW, Australia
Nagoya, Japan
Beijing, China
April 2014

Quan Bai
Fenghui Ren
Minjie Zhang
Takayuki Ito
Xijin Tang

Contents

HIPRank: Ranking Nodes by Influence Propagation Based on Authority and Hub

Wen Zhang, Song Wang, Guangle Han, Ye Yang, and Qing Wang

Abstract Traditional centrality measures such as degree, betweenness, closeness and eigenvector ignore the intrinsic impacts of a node on other nodes. This paper proposes a new algorithm, called HIPRank, to rank nodes based on their influences in the network. HIPRank includes two sub-procedures: one is to predefine the importance of an arbitrary small number of nodes with users' preferences, and the other one is to propagate the influences of nodes with respect to authority and hub to other nodes based on HIP propagation model. Experiments on DBLP citation network (over 1.5 million nodes and 2.1 million edges) demonstrate that on the one hand, HIPRank can prioritize the nodes having close relation to the user-preferred nodes with higher ranking than other nodes, and on the other hand, HIPRank can retrieve the authoritative nodes (with authority) and directive nodes (with hub) from the network according to users' preferences.

Keywords Influence propagation • Network modeling • Nodes ranking

1 Introduction

Recently, network-based search arises in both research and application areas. Those traditional algorithms, such as Google's PageRank [8] and Kleinberg's HITS [5], assumed a global view on the structure of the network to treat all the users' preferences [4] in ranking nodes equally. However, in most cases, this

W. Zhang (✉)
School of Management and Economics, Beijing University of Chemical Technology,
Beijing, China
e-mail: zhangwen@mail.buct.edu.cn

S. Wang
Chinese Academy of Sciences, University of Chinese Academy of Sciences, China
e-mail: wangsong@nfs.iscas.ac.cn

G. Han • Y. Yang • Q. Wang
Chinese Academy of Sciences, Beijing, China
e-mail: guangle@nfs.iscas.ac.cn; yangye@nfs.iscas.ac.cn; wq@nfs.iscas.ac.cn

© Springer Japan 2015
Q. Bai et al. (eds.), *Smart Modeling and Simulation for Complex Systems*, Studies
in Computational Intelligence 564, DOI 10.1007/978-4-431-55209-3_1

assumption may be inappropriate because of the actual differences of users' personal preferences in nodes ranking. Thus, PPR (Personalized PageRank) [1, 2, 4] has been proposed to solve the problem of personal preferences. Although most of the research endeavor has been invested in speeding up the PPR to make its computation practical [1, 2, 4], similar to Pei Li et al [6], we attempted a different treatment to considering users' personal preferences by propagating the user-predefined importance of nodes in the network. Inspired by the Hyperlink-Induced Topic Search (HITS) [5], which measures the importance of web pages in the web network using *authority* and *hub*, we believe that connected nodes in the network can influence each other and the influence of nodes should be propagated bidirectionally (forward and backward) rather than unidirectional propagation in Pei Li et al [6].

The main contributions of this paper can be summarized in two aspects. First, we argue that the user preference or prior knowledge of nodes should be taken into consideration when ranking nodes in the network. Second, we propose a new influence propagation model called HIP to describe the bidirectional propagation of predefined importance of *authority* and *hub* over nodes accompanied with random walk paths. Based on these two aspects, we propose a new algorithm called HIPRank to rank individuals in the network. The computation complexity of our proposed algorithm HIPRank is much less than those proposed to solve the PPR problem.

The remainder of this paper is organized as follows. Section 2 describes the motivation. Section 3 presents the related work. Section 4 proposes the HIP model and HIPRank algorithm for ranking nodes in the network. Section 5 conducts experiments. Section 5.3 concludes the paper. Table 1 lists the notations used in this paper.

Table 1 Notations used in this paper

$G(V, E)$	G is a directed graph; V is the set of vertices of G; E is the set of edges of G
Z_a	The vector of predefined *authority* importance of all nodes
Z_h	The vector of predefined *hub* importance of all nodes
$Z_{a,i}(b)$	The influence of *authority* received by node b on the i-th propagation step
$Z_{h,i}(b)$	The influence of *hub* received by node b on the i-th propagation step
$Z_a(p, q)$	The influence of *authority* propagated from node p to q through all the paths from node p to q
$Z_h(p, q)$	The influence of *hub* propagated from node p to q through all the paths from node p to q
K	The maximum propagation step in HIP model
N	The nodes size of a directed graph
M	The edges size of a directed graph
P	The number of edges traversed in the propagation process
R_a	The *authority* ranking vector of all nodes
R_h	The *hub* ranking vector of all nodes
$R_{a,i}$	The *authority* ranking vector of all nodes on the i-th iteration/step
$R_{h,i}$	The *hub* ranking vector of all nodes on the i-th iteration/step
$O(i)$	The out-going neighbors of node i

2 The Motivation

Let W be the adjacency matrix of a directed graph G, and T is the transpose of W. For the element at p-th row and q-th column of W, in HITS, $W_{p,q} = 0$ if $(p, q) \notin E$, otherwise $W_{p,q} = 1$. In HIPRank, we normalize W and T, each row of W and T is normalized to one unless all elements in this row are zero, W' and T' denote the normalized W and T respectively. For the p-th row and q-th column element of W', $W'_{p,q} = 0$ if $(p, q) \notin E$, otherwise $W'_{p,q} = w(p,q)/\sum_{i \in O(p)} w(p,i)$, where $w(p,q)$ is the weight of edge (p,q) in W, the same rule to T'.

For the network depicted in Fig. 1, using HITS to rank nodes in it, the normalized *authority vector* of the nodes is [0.202,0.095,0.335,0.166,0.202] and the normalized *hub vector* of the nodes is [0.190,0.240,0.165,0.240,0.165]. HITS does not consider the initial importance of the nodes. That is, the initialized *authority* and *hub* are not propagated in HITS algorithm to rank nodes. The basic idea behind HITS is to measure the centrality (importance) of web pages in the web network using *authority* and *hub*, *authority* estimates the value of the content of a web page, and *hub* estimates the value of its links to other pages. The matrix form of HITS can be formulated as below in Eq. (1).

$$R_{a,k} = c_k R_{h,k-1} W; R_{h,k} = c'_k R_{a,k-1} T \tag{1}$$

Here, k represents the k-th iteration, c_k and c'_k are normalization factors. Based on Eq. (1), the *authority* and *hub* scores of HITS can be computed iteratively, and this process is proved convergent if we normalize the two score vectors of *authority* and *hub* after each iteration [5].

When considering a case that a user has some prior knowledge or preference on some nodes in Fig. 1 and prefers to retrieve nodes like node b, thus the initial importance, which represents the user's preference, may cast a crucial influence on the result of the ranking. However, in this case, typical HITS algorithm will not work because HITS does not take the initial importance of nodes into account when

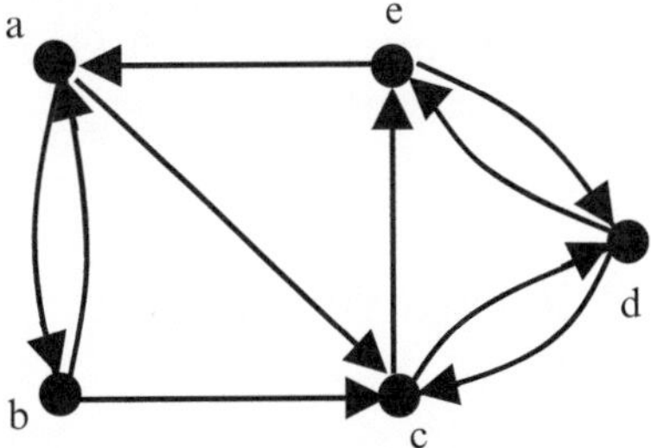

Fig. 1 A directed network consists of 5 nodes and every node has predefined importance in two aspects, including *authority* and *hub*

Table 2 Normalized HITS scores and HIPRank scores

$Z_a \& Z_h$	Normalized HIPRank Scores (%)
$Z_a = [0.2, \mathbf{0}, 0.2, 0.2, 0.2]$	$Authority : [20.63, 8.75, 31.97, 18.02, 20.63]$
$Z_h = [0.2, 0.2, 0.2, 0.2, 0.2]$	$Hub : [17.92, 23.43, 17.61, 23.43, 17.61]$
$Z_a = [0.2, \mathbf{0.2}, 0.2, 0.2, 0.2]$	$Authority : [20.38, 9.23, 32.58, 17.41, 20.38]$
$Z_h = [0.2, 0.2, 0.2, 0.2, 0.2]$	$Hub : [18.58, 23.58, 17.13, 23.58, 17.13]$
$Z_a = [0.2, \mathbf{0.4}, 0.2, 0.2, 0.2]$	$Authority : [20.16, 9.68, 33.14, 16.86, 20.16]$
$Z_h = [0.2, 0.2, 0.2, 0.2, 0.2]$	$Hub : [19.20, 23.72, 16.68, 23.72, 16.68]$
HITS scores	$Authority : [20.2, 9.5, 33.5, 16.6, 20.2]$
	$Hub : [19.0, 24.0, 16.5, 24.0, 16.5]$

The bold values indicate the authority score of node b

ranking nodes in the network. Additionally, HITS is originally designed to rank the web network. Simply applying it to rank nodes in general network may result in unexpected results, because the "random jumping" behavior of web network is not suitable for modeling some friendship based social networks.

To solve this problem, IPRank [6] is proposed to consider the initialized importance of nodes in the network. However, in other cases, predefining importance of all the nodes in only one dimension is not enough. For instance, in a paper citation network, some papers, such as surveys and reviews, cite a large number of other papers because they focus on reviewing the recent advancements in a domain. We may call this kind of papers *hub paper*. Meanwhile, other papers have a large number of citations because they focus on presenting specialized algorithms or pioneering approaches to some difficult problems in a domain. We may call this kind of papers *authority paper*. Simply using only one dimension to measure the predefined importance of these two kinds of papers in a citation network will cause a great loss of important information.

Motivated by the problems of HITS and IPRank, we propose a new influence propagation model called HIP to model bidirectional propagation of influences in the network, and a new ranking algorithm called HIPRank to rank nodes, based on the global structural contexts of the network accompanied with predefined importance of authority and hub. Table 2 shows the normalized HITS scores and HIPRank scores corresponding to different predefined importance Z_a and Z_h. The decay function is $f(k) = 0.8^k$.

3　Related Work

The related work of this paper can be categorized into two aspects. One is personalized PageRank. The basic idea is that while the global network topology inducing the adjacent matrix in PageRank is the same for all users, the preference vectors inducing users' preference on nodes are different for different users. However, a difficult problem of personalized PageRank is it needs huge computation. To address this problem, many solutions were proposed such as probabilistic random walk with external memory indexing [1], incremental computation [4] and top-K search with bounded accuracy [2]. Actually, in most cases, user preference is dependent on different domains and not a constant. Thus, it is very hard to capture user preference correctly due to its diversity and volatility.

This problem brings about the other related work of the paper, i.e. influence-propagation-based ranking methods. The basic idea is that for a search task in a given domain, a user has some prior knowledge of the influential nodes in the network. Thus, by propagating influence to other nodes based on the network topology, all the nodes in the work obtain their importance ranking. The methods in this aspect include IPRank with propagating influence based on PageRank [6], propagating relevance and irrelevance [9], propagating trust and distrust [3], etc. HIPRank also ranks nodes based on influence propagation. However, we are different from the previous work, we propagate *authority* and *hub* of nodes in the network by introducing the idea of HITS algorithm [5].

4 HIPRank

4.1 HIP Propagation Model

Let Z_a and Z_h be two vectors to represent predefined importance of *authority* and *hub* of nodes in G respectively, while Z_a represents the initialized *authority* values of nodes and Z_h represents the initialized *hub* values of nodes, and all the elements in Z_a and Z_h are non-negative.

In HITS algorithm, the *authority* of node c comes from the *hub* of its in-neighbors within 1-step hop and the *hub* of node c comes from the *authority* of its out-neighbors within 1-step hop. In HIP, the influence propagated bidirectionally. The *authority* of node c comes not only from the *hub* of its in-neighbors within 1-step hop, but also from the *hub* of its in-neighbors within k-step hop ($1 < k < K$, K is predefined as the maximum propagation step). The *hub* of node c comes not only from the *authority* of its out-neighbors within 1-step hop, but also from the *authority* of its out-neighbors within k-step hop ($1 < k < K$). For instance, assuming there is a path $p = < v_0, v_1, v_2, \ldots, v_k >$, the *hub* $Z_h(0)$ propagating from v_0 to v_k in forward direction, contributes to the *authority* of v_k. Equation (2) shows the received influence of v_k from v_0.

$$Z_a(0, k) = Z_h(0) \cdot f(k) \cdot \prod_{i=0}^{k-1} W'_{i,i+1} \qquad (2)$$

The authority $Z_a(k)$ propagating from v_k to v_0 in backward direction, contributes to the *hub* of v_0. Equation (3) shows the received influence of v_0 from v_k.

$$Z_h(k,0) = Z_a(k) \cdot f(k) \cdot \prod_{i=0}^{k-1} T'_{i,i+1} \tag{3}$$

We introduce a discrete *decay function* $f(k) = c^k$ to capture the retained influence on the k-th step hop. Here $k \in \{1, 2, 3, \ldots, K\}$, K is predefined as the maximum propagation step, and $0 < c < 1$. Generally, $f(k) < 1$, and the lager k results in smaller $f(k)$. In order to decide the maximum propagation step K, a threshold h that satisfies the following condition needs to be specified.

$$f(K) \geq h \text{ and } f(K+1) < h \tag{4}$$

Proposition 1. *Without decay function, that means when $f(k)$ is a constant, influence also decays in HIP model.*

Proof. According to Eqs. (2) and (3), when $f(k)$ is a constant, here denoted by C, the *authority* influence propagating from v_0 to v_k can be calculated by $Z_a(0, k) = Z_h(0) \cdot C \cdot \prod_{i=0}^{k-1} W'_{i,i+1}$, since each row in W' is normalized to one, so each element in W' is less than 1 when $k \to \infty$ $\prod_{i=0}^{k-1} W'_{i,i+1} \to 0$, thus the propagated influence decays via the propagation path. Similarly, the same with *hub* influence. Proposition 1 holds.

The reason why we introduce decay function $f(k)$ in HIP model is that in HIP the decay of influence is determined by the topological structure of the network, which is uncontrolled by users. With a specified decay function, a user can control the propagation of influence over nodes.

4.2   HIPRanking Nodes

Based on the HIP propagation model, the HIPRank scores of nodes in the network can be defined as follows.

Definition 1. The HIPRank scores of a node in the network consist of a *authority* score and a *hub* score, and both scores are measured by the initialized importance of this node and the influence propagated to this node from other nodes.

The basic idea behind HIPRank is that, the more influence of *authority* or *hub* a node receives from other nodes, the more authoritative or directive the node is in the network. Different from IPRank [6], the influence propagation in HIPRank considers initial importance of a node in two dimensions: *authority* and *hub*. The initial *authority* and *hub* of nodes are represented by Z_a and Z_h respectively. Both Z_a and Z_h are propagated in HIP to impact *authority* and *hub* scores of nodes as shown in Table 2.

The *authority* and *hub* of nodes in HIPRank can be computed in the same manner. The only difference between *authority* and *hub* lies in that, the *authority* of a node propagates in the backward direction, and the *hub* of a node propagates in the forward direction. For this reason, we use the computation of *authority* score of a node in HIPRank as an example to show how the HIPRank algorithm ranks nodes in the network.

With the decay during the propagation, the propagated influence of *hub* and *authority* can be ignored after K steps. Therefore, we only need to collect the propagated influences that reach a node within K steps with random walk [7, 10].

Considering the node a in the graph G in Fig. 1 and supposing $K = 1$, that is, the influence of *authority* and *hub* propagates within 1 step. To compute the *authority* score of node a, we reverse all edges and traverse b and e starting from the node a. And two random walk paths (b, a) and (e, a) in G that reach node a within 1 step hop are collected. Thus, the 1 step *authority* propagation from node b and e to a as $Z_{a,1}(a)$ can be denoted in Eq. (5), where $Z_h(b)$ and $Z_h(e)$ represent the hub scores of node b and e, respectively.

$$Z_{a,1}(a) = f(1) \cdot (Z_h(b)W'_{b,a} + Z_h(e)W'_{e,a}) \tag{5}$$

The *authority* score of node a in Fig. 1 can be calculated in Eq. (6), where $Z_a(a)$ represents the initial *authority* score of node a.

$$R_a(a) = Z_{a,1}(a) + Z_a(a) \tag{6}$$

Note that, different random paths generated by HIPRank may have common sub-paths, which could be reused to save computational cost. For example, *hub* propagation along paths $< c, d, c, e >$ and $< c, d, c >$ are generated by HIPRank queries for node e and c, and these two paths share the common sub-path $< c, d, c >$, which can be reused in practical computation.

For all the nodes in a network, we develop an algorithm, called *HIPRank-All* to compute their HIPRank scores in matrix form as shown in Algorithm 1.

In HIPRank, the initial *authority* and *hub* of all nodes are stored in two vectors Z_a and Z_h respectively. Taking the computation of *authority* score as an example, in the first step, all the nodes propagate *hub* to their out-neighbors with decay factor $f(1)$. Considering the *hub* received by a node which contributes to its *authority*, let $I_1(v)$ be the in-neighbor set of node v within 1 step hop. The *hub* received by v is $Z_{a,1}(v) = f(1) \cdot \sum_{n=1}^{I_1(v)} Z_h(n)W'_{n,v}$. Further, considering all the nodes in the network, we obtain $Z_{a,1} = f(1) \cdot Z_h W'$ in matrix form. In the second step, all the nodes that are 2-step in-neighbors to node v, represented by $I_2(v)$, will also propagate *hub* to v and contribute to the *authority* of v. Thus, the *authority* of v obtained from the second step is $Z_{a,2}(v) = f(2) \cdot \sum_{n=1}^{I_2(v)} Z_h(n)W'^2_{n,v}$. Considering all the nodes in the network, we obtain $Z_{a,2}(v) = f(2) \cdot Z_h W'^2_{n,v}$ in matrix form. By analogy, the obtained *authority* vector of all nodes in the network on the k-th step can be computed by Eq. (7).

$$Z_{a,k} = f(k) \cdot Z_h W'^k \tag{7}$$

Algorithm 1 HIPRank-All(G, Z_a, Z_h, h, W, T)

Input: Graph $G(V, E)$, initial *authority* vector Z_a, initial *hub* vector Z_h, threshold h for deciding
    the maximum propagation step, W is the adjacency matrix of G, and T is the transpose of W.
Output: HIPRank scores R_a and R_h
1: initialize $R_a = Z_a$; $R_h = Z_h$;
2: obtain W' and T' by normalizing W and T;
3: **for** every node $v \in V$ **do**
4:     obtain K according to Equation (4);
5:     AuthorityRecursion$(v, Z_h(v), 0, K)$;
6:     HubRecursion$(v, Z_a(v), 0, K)$;
7: **end for**
8: **return** R_a and R_h;
9: **procedure** AUTHORITYRECURSION(v, x, y, K)
10:    $y = y + 1$;
11:    **for** every node u in out-neighbor set of node v **do**
12:       $R(u) = R(u) + x \cdot W'_{v,u} \cdot f(y)$
13:       **if** $y < K$ **then**
14:          AuthorityRecursion$(u, x \cdot W'_{v,u}, y, K)$;
15:       **end if**
16:    **end for**
17: **end procedure**
18: **procedure** HUBRECURSION(v, x, y, K)
19:    $y = y + 1$;
20:    **for** every node u in out-neighbor set of node v **do**
21:       $R(u) = R(u) + x \cdot T'_{v,u} \cdot f(y)$
22:       **if** $y < K$ **then**
23:          HubRecursion$(u, x \cdot T'_{v,u}, y, K)$;
24:       **end if**
25:    **end for**
26: **end procedure**

As a result from Definition 1, in HIPRank, the *authority* ranking vector obtained within k steps can be described in Equation (8).

$$R_{a,k} = \sum_{i=1}^{k} Z_{a,i} + Z_a = \sum_{i=1}^{k} f(i) \cdot Z_h W'^i + Z_a \qquad (8)$$

In the same manner, we can obtain the *hub* ranking vector within k steps, as described in Eq. (9).

$$R_{h,k} = \sum_{i=1}^{k} Z_{h,i} + Z_h = \sum_{i=1}^{k} f(i) \cdot Z_a T'^i + Z_h \qquad (9)$$

Equations (8) and (9) describe the main computation of *HIPRank-All* algorithm. The time complexity of *HIPRank-All* algorithm is $O(KP)$, P is the number of edges traversed in the propagation process. One useful proposition about HIPRank computing is given below.

Proposition 2. *When $f(k) = c^k\,(0 < c < 1)$ and $k \to \infty$, HIPRank is convergent.*

Proof. According to Eq. (8), since each row in W'^k is normalized to one, when $k \to \infty$, $W'^k \to 0$. The decay function is $f(k) = c^k\,(0 < c < 1)$, therefore:
$$R_{a,k} = Z_h\big(cW' + c^2 W'^2 + c^3 W'^3 + \ldots + c^k W'^k\big) + Z_a$$

$$k \to \infty \quad R_a = Z_h(\mathbf{E} - cW')^{-1} + Z_a \tag{10}$$

Similarly:

$$R_h = Z_a(E - cT')^{-1} + Z_h \tag{11}$$

Here, $\mathbf{E}$ is the identity matrix and $\mathbf{0}$ is the zero matrix. Equations (10) and (11) show that when $f(k) = c^k\,(0 < c < 1)$ and $k \to \infty$, R_a and R_h are convergent. Proposition 2 holds.

In HITS the only factor that influences the final *authority* and *hub* scores is the topological structure of the network, while in HIPRank, from Eqs. (10) and (11), we can see that both the topological structure of the network and the initialized importance of *authority* and *hub* can influence the final ranking scores.

5 Experiments

5.1 *Appropriate Propagation Step K*

For obtaining acceptable scores for all the nodes, an appropriate propagation step K should be specified. We conduct a simulation on a PC with a 3.4 GHz CPU and an 8GB RAM to exam how to set an appropriate K. We use three random networks called G_1(1 million nodes and 3 million edges), G_2(1 million nodes and 5 million edges) and G_3(1 million nodes and 10 million nodes). Similar to Pei Li et al [6], we also use the precision defined by average $R_k(a)/R(a)$ to observe the convergence rate of HIPRank on those three graphs. For G_1 we set the $|E|/|V| = 3$, Fig. 2a shows that the error of precision is below 0.01 after 9 iterations; for

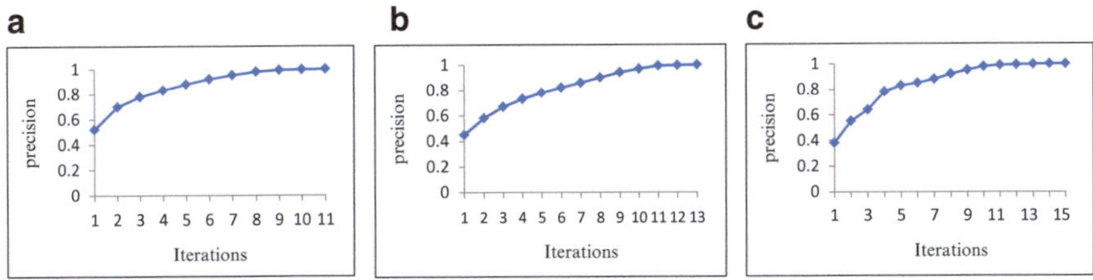

Fig. 2 Convergence rate of three networks (left to right G_1, G_2 and G_3)

G_2, $|E|/|V| = 5$, results in Fig. 2b shows after 11 iterations the error of precision is below 0.01; for G_3, $|E|/|V| = 10$, Fig. 2c shows that the error of precision is below 0.01 after 12 iterations. So, with the proportion of the number of edges and nodes increasing, HIPRank needs more iteration to obtain high convergence rate. In Fig. 2 when $K = 10$, the precision of G_1 is 0.996, G_2 is 0.97 and G_3 is 0.95, so we recommend when $|E|/|V| < 10$, the max propagation step K is set to 10 and when $|E|/|V| \geq 10$, the K should be bigger than 10.

5.2 Time Cost of HIPRank

The time complexity of *HIPRank-All* algorithm is $O(KP)$, where K is the maximum propagation steps and P is the number of edges traversed in the propagation process. Figure 3a shows the time cost when K increases. We can see that when K is smaller than 9, which means the influence of a node propagates less than 9 steps, the time cost of HIPRank keeps very small because the number of traversed edges in the propagation process would be not very large at this duration. However, when K is larger than 10, there is a linear increase in the time cost as we explain that all the edges in the network are traversed in the propagation process. Figure 3b shows the time cost when M increases. We set K as 10 to obtain high convergence rate. As M increases, P increases at the same time, and the time cost of HIPRank increases linearly when M increase from 0.1 to 0.7 million.

In practice, P is usually much larger than K. Thus, the computation complexity is almost decided by P, which is not larger than the number of edges (M) in the work. In this case, traversed edges in the proposed HIP rank model are much smaller than that in models proposed by [2] and [9] to find the top-K relevant nodes.

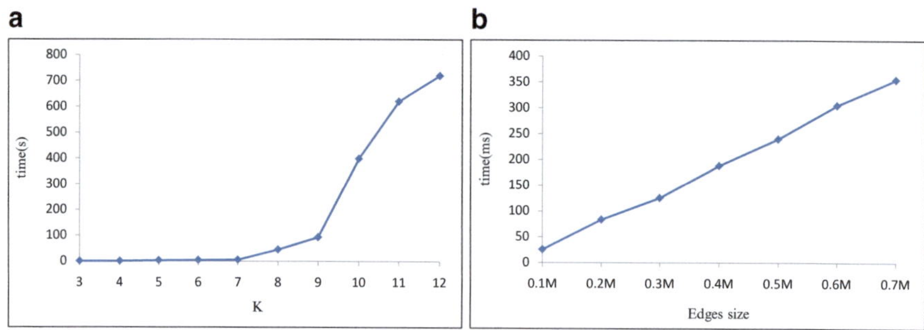

Fig. 3 (**a**) Time cost when K increases ($N = 1.5$ million, $M = 3.5$ million). (**b**) Time cost when edges size increases ($K = 10$, $N = 0.1$ million)

5.3   Results on DBLP Paper Citation Network

We build a large paper citation network using the citation information of the entire DBLP conference papers. This network consists of 1,511,035 papers (nodes) and 2,084,019 citations (edges).

Three expected outcomes of the experiment are: (i) if a user has already known that some papers are important in some fields, for example, we know that the paper "C4.5: Programs for Machine Learning" is an authoritative paper in KDD area, a user can find other authoritative and directory papers (*authority paper* and *hub paper*) in KDD area using the HIPRank model by setting high *authority* scores to this paper; (ii) Papers with high *authority* scores have larger possibilities to propose novel algorithms or pioneering methods to solve difficult problems in a domain. We can call these papers *authority paper*; (iii) Papers with high *hub* scores have larger possibilities to be surveys, overviews, or reviews, which called *hub paper*. The decay function is $f(k) = 0.8^k$ and the maximum propagation step is set to 10. The initial *authority* and *hub* values of the predefined nodes are set to $N/2$, the initial values of the rest nodes are set to $1/N$, and then we normalize all the initial values to the range from 0 to 1 for all the nodes, where N is the number of nodes in the network.

First we use HITS on the paper citation network, and the corresponding top-10 *authority paper* and top-10 *hub paper* are shown in Table 3a. For the top-10 *authority paper*, papers rank high only because they have high citations; for the top-10 *hub paper*, papers rank high mainly because they cite *authority paper*.

Second, we bias the ranking to a special area by predefining importance (both *authority* and *hub*) for papers. In Table 3b, papers published in SE (Software Engineering) area conferences (here we use ICSE, FSE, ESEM, and SIGSOFT) are given higher predefined *authority* and *hub* scores. Then we obtain top-10 *authority paper* and top-10 *hub paper* in Software Engineering area.

Third, we bias HIPRank to KDD(Knowledge Discovery and Data Mining) area (here we use conferences: SIGKDD, PAKDD, PKDD, and ICDM) and show results in Table 3c. Those *authority paper* and *hub paper* of special areas listed in Table 3b,c show that our HIPRank with predefined *authority* and *hub* scores produces reasonable results.

From Table 3b,c, we can see that many papers which are not initialized with relatively high *authority* and *hub* are retrieved from DBLP dataset, such as those papers from ACM TOSEM for SE area and SGIMOD for KDD area. These outcomes illustrate that HIPRank is enlightening in discovering papers by user preference.

Table 3 (**a**) Results of HITS. (**b**) HIPRank on SE area. (**c**) HIPRank on KDD area

(a)

Top10 *Authority papers*	Conf	Top10 *Hub papers*	Conf
C4.5: Programs for Machine Learning	BOOK	Data Mining: An Overview from a Database Perspective	IEEE Trans
Fast Algorithms for Mining Association Rules in Large Databases	VLDB	Scalable Algorithms for Mining Large Databases	SIGKDD
Mining Association Rules between Sets of Items in Large Databases	SIGMOD	Mining Query Logs: Turning Search Usage Data into Knowledge	FTIR
Introduction to Algorithms	BOOK	Scalable frequent-pattern mining methods: an overview	SIGKDD
Introduction to Modern Information Retrieval	BOOK	ACIRD: Intelligent Internet Document Organization and Retrieval	IEEE Trans
Modern Information Retrieval	BOOK	ART: A Hybrid Classification Model	Machine Learning
Induction of Decision Trees	BOOK	Ratio Rules: A New Paradigm for Fast, Quantifiable Data Mining	VLDB
Compilers: Princiles, Techniques, and Tools	BOOK	Using Information Retrieval techniques for supporting data mining	Data Knowledge Engineering
The Anatomy of a Large-Scale Hypertextual Web Search Engine	WWW	Association Rule Mining, Models and Algorithms	BOOK
Mining Frequent Patterns without Candidate Generation	SIGMOD	From intra-transaction to generalized inter-transaction: Landscaping multidimensional contexts in association rule mining	Information Sciences

(b)

The Model Checker SPIN	IEEE Trans	Research Directions in Requirements Engineering	FOSE
Communicating Sequential Processes	Communications of the ACM	A brief survey of program slicing	SIGSOFT
Statecharts: A Visual Formalism for Complex Systems	Science of computer programming	Architecture Reconstruction	SE
Experiments of the Effectiveness of Dataflow and Controlflow-Based Test Adequacy Criteria	ICSE	Requirements interaction management	CSUR
Compilers: Princiles, Techniques, and Tools	BOOK	A schema for interprocedural modification side-effect analysis with pointer aliasing	TOPLAS

(continued)

Table 3 (continued)

Top10 *Authority papers*	Conf	Top10 *Hub papers*	Conf
A Formal Basis for Architectural Connection	TOSEM	Software Unit Test Coverage and Adequacy	CSUR
Software Processes Are Software Too	ICSE	Context-aware statistical debugging: from bug predictors to faulty control flow paths	ASE
Automatic Verification of Finite-State Concurrent Systems Using Temporal Logic Specifications	TOSEM	The IBM-McGill project on software process	CASCON
Bandera: extracting finite-state models from Java source code	SE	Profile-guided program simplification for effective testing and analysis	SIGSOFT
(c)			
Fast Algorithms for Mining Association Rules in Large Databases	VLDB	Scalable frequent-pattern mining methods: an overview	SIGKDD
C4.5: Programs for Machine Learning	BOOK	Scalable Algorithms for Mining Large Databases	SIGKDD
Mining Association Rules between Sets of Items in Large Databases	SIGMOD	Data Mining: An Overview from a Database Perspective	IEEE Trans
Mining Frequent Patterns without Candidate Generation	SIGMOD	From intra-transaction to generalized inter-transaction: Landscaping multidimensional contexts in association rule mining	Information Sciences
Mining Sequential Patterns	BOOK	Association Rule Mining, Models and Algorithms	BOOK
Induction of Decision Trees	BOOK	Off to new shores: conceptual knowledge discovery and processing	Human Computer Studies
An Efficient Algorithm for Mining Association Rules in Large Databases	VLDB	A template model for multidimensional inter-transactional association rules	VLDB
An Effective Hash Based Algorithm for Mining Association Rules	SIGMOD	Efficient dynamic mining of constrained frequent sets	TODS
Dynamic Itemset Counting and Implication Rules for Market Basket Data	SIGMOD	Mining Frequent Patterns without Candidate Generation: A Frequent-Pattern Tree Approach	Data Mining Knowledge Discovery
Mining Quantitative Association Rules in Large Relational Tables	SIGMOD	ART: A Hybrid Classification Model	Machine Learning

Conclusion

This paper proposes a new ranking model, called HIPRank, to rank individuals based on their influence propagation of *authority* and *hub* in the network. The basic idea of HIPRank is to make use of user preference and prior knowledge of nodes to initialize the *authority* and *hub* of nodes. Then, the initialized *authority* and *hub* of each node are propagated to other nodes through the topology of the network. Finally, the importance of each node in the network is measured by summing their initialized *authority* and *hub* with the propagated *authority* and *hub* from other nodes within the predefined K-step hops. Also, users can control the propagation by defining decay function and the maximum propagation step K. Experiments on synthetic data and the real DBLP citation dataset demonstrate the effectiveness of the proposed approach in retrieving user-intended authoritative and directory individuals from the network.

Acknowledgements This research was supported in part by National Natural Science Foundation of China under Grant Nos. 61073044, 71101138, 61003028 and 61379046; National Science and Technology Major Project under Grant Nos. 2012ZX01039-004; Beijing Natural Science Fund under Grant No.4122087; State Key Laboratory of Software Engineering of Wuhan University.

References

1. Csalogny, K., et al.: Towards scaling fully personalized pagerank: algorithms, lower bounds, and experiments. Internet Math. **2**(3):333–358 (2005)
2. Fujiwara, Y., et al.: Efficient personalized pagerank with accuracy assurance. In: Proceedings of the 18th ACM SIGKDD International Conference on Knowledge Discovery and Data Mining, pp. 15–23 (2012)
3. Guha, R.V., et al.: Propagation of trust and distrust. In: Proceedings of the 13th International Conference on World Wide Web, pp. 403–412 (2004)
4. Jeh, G., Widom, J.: Scaling personalized web search. In: Proceedings of the 12th International Conference on World Wide Web, pp. 271–279 (2003)
5. Kleinberg, J.M.: Authoritative sources in a hyperlinked environment. J. ACM. **46**(5):604–632 (1999)
6. Li, P., et al.: Ranking individuals and groups by influence propagation. In: Proceedings of the 15th Pacific-Asia Conference on Advances in Knowledge Discovery and Data Mining, pp. 407–419 (2011)
7. Motwani, R., Raghavan, P.: Randomized Algorithms. Cambridge University Press, Cambridge (1995)
8. Page, L., Brin, S., Motwani, R., Winograd, T.: The pagerank citation ranking: bringing or-der to the web. Technical Report 1999-66, Stanford InfoLab (1999)
9. Sarkar, P., Moore, A.W.: Fast dynamic reranking in large graphs. In: Proceedings of the 18th International Conference on World Wide Web, pp. 31–40 (2009)
10. Valente, T.: Network Models of the Diffusion of Innovations. Hampton Press, Cresskill (1995)

Benefits of Generalised Microsimulation

Daniel Keep, Ian Piper, and Anthony Green

Abstract This paper states our position regarding the desirable properties of a discrete simulation environment and details our response to this. Following a brief introductory examination of the pre-existing art in microsimulation modelling, we describe our approach and detail its structure and use.

Keywords Microsimulation • Microsimulation modelling • Simulation

1 Introduction

One of the notable techniques in discrete event modelling is microsimulation modelling (MSM). Unfortunately, MSM is typically discussed in terms of specific application domains, rather than as a general technique in itself. We believe that this has led to the potential of MSM, as a tool for complex multi-domain simulations, to be largely overlooked.

Although there has been some work towards such generalisation [16], its focus has been largely on tools for the development of simulation programs. We are interested in building a framework that allows for the easy reuse and composition of simulations through the use of components. In addition, it is our contention that it should be possible for non-trivial simulations to be constructed by non-programmers.

We are also interested in combining the best aspects of various simulation techniques, using our MSM constructs as a common base. This, we believe, facilitates the creation of integrated systems for modelling of real-world problems involving multiple phenomena and processes.

We believe these aims are feasible; the existence and demonstrated utility of our MSM framework, simulation modules and associated tools confirm this position, as well as providing a useful base for further research and development.

D. Keep • I. Piper (✉) • A. Green
University of Wollongong, Wollongong, NSW, Australia
e-mail: ian@uow.edu.au

© Springer Japan 2015
Q. Bai et al. (eds.), *Smart Modeling and Simulation for Complex Systems*, Studies in Computational Intelligence 564, DOI 10.1007/978-4-431-55209-3_2

2   Microsimulation Modelling

Microsimulation modelling was first described by Orcutt in 1957. In MSM, the primary modelling construct is the individual [10]. The population is modelled either by a representative sample, or by a complete census. In both cases, each member of the population has its own state and behaviour. This is in contrast to the standard modelling techniques of the day which aggregated the state and behaviour of the individuals which made up the population.

MSM techniques are applicable to a wide range of problem domains ranging from epidemiology [3] to transportation [19]. However, there appears to be an unfortunate level of "ghettoisation", in that references to MSM are almost always made in the context of a particular area of application. There are many examples of *transport* microsimulation, *social* microsimulation, or *economic* microsimulation, but comparatively few examples of *general* microsimulation.

2.1   On Agents and MSM

Spielauer notes that agent-based simulation can be considered a kind of dynamic MSM, but generally is not [15]. He states that both techniques developed almost entirely in isolation from one another.

Crooks and Heppenstall state that the crucial difference between an agent-based approach and MSM is that MSM does not model the impact individuals can have on the model's policy, nor does it model interactions between individuals [5]. This is not a hard-and-fast rule; existing work shows how elements of agent-based modelling can be introduced into an MSM context [8, 20].

It is interesting to note that Stroud et al. refer to EpiSimS (which is arguably a dynamic microsimulation) as an agent-based simulation [17], whilst Chen refers to the same approach variously as "microsimulation", "agent-based simulation" and "agent-based microsimulation" [4]. In practice, there is much common ground between the two approaches.

2.2   Observations on the State of the Art

We are primarily interested in modelling the behaviour and response of individuals. Both MSM and Agent-Based Modelling (ABM) were developed to facilitate such modelling. In practice, MSMs are built for specific purposes. For example, TRANSIMS is a transport microsimulation [14]. Even if you are using a framework such as ModGen or UMDBS [13], the end result is a self-contained, stand-alone simulation program with a specific, well-defined purpose.

It is our belief that a greater emphasis on the development of generalised microsimulation technologies would be of great benefit, both in domains already served by existing tools and in domains as yet untapped. Although there is unquestionable value in building simulations for specific areas of interest, there is also a need for simulations which cover multiple problem domains. An example of this is the work of Perez et al. at the University of Wollongong in which the interaction between population changes and the transportation network are being analysed. Despite this being a natural problem to examine, a significant portion of their work has been spent on interfacing the two major simulations components. This would also serve to combat the ghettoisation of MSM.

There appears to be a general perception that models which involve behaviour necessitate the use of "full-blown" ABM techniques. We believe that a considerable (and very useful) degree of expressivity can be achieved by introducing simple, limited aspects of ABM to MSM, rather than abandoning MSM in favour of ABM. Birkin and Wu would appear to share this belief, stating: ". . .it is more constructive to view the relationship of ABM to MSM as one of complementarity rather than supremacy." [2]

2.3 Methods for Developing a New Simulation

There are three broad strategies for constructing a simulation. First, the simulation can be created from scratch. Secondly, the simulation can be built on a simulation framework. Thirdly, the simulation can be assembled using one or more simulation components.[1] Of these, we believe the third to be the most interesting and attractive.

Creating simulations from scratch naturally involves a significant amount of new design and coding effort. Some of this work can be saved by modifying or combining existing simulations. This still requires a significant amount of time and effort, in some cases potentially more than would be required to write it from scratch. This is particularly true if combining simulations with substantially different data formats, interfaces, approaches to time and space, etc.

Frameworks significantly reduce this burden by implementing common pieces of functionality. If implementing an entirely novel model, this is probably the most attractive approach. However, most frameworks are designed to produce stand-alone, single-purpose simulations, which cannot be readily re-purposed or combined. Thus, there may still be significant barriers to combining disparate simulations, even within the same framework.

Our approach is to design the framework not just to implement commonly required support code, but to also ensure that multiple, independent processes can co-exist within the same simulation without requiring code changes at any level.

---

[1] Ideally, these components would be pre-existing, although additional ones could be written with no more effort than required in the second approach.

Once a particular capacity has been implemented atop the framework, it is available for use alone or in combination with any other such components, without additional code. The desired traits of reusability and composability are inherent in this strategy, the value of which have been previously recognised, at the conceptual level, by Balci et al. [1]. This has been borne out in practice, with the non-programming members of our team able to construct new, complex simulations without the need for a programmer.

3   Our Work

The work of our group is focused on developing and applying a generalised approach to the construction of MSM simulations. This has resulted in the creation of the *Simulacron* MSM framework (and supporting code), a number of simulation modules, and associated tools, as well as the design and construction of a number of simulations which make use of the framework, modules and tools. Further details on these can be found in [9]. The guiding principle behind the design of this software has been "keep it simple; if you can't, keep it general."

3.1   *Simulacron Framework*

Conceptually, the framework's most important task is defining the set of fundamental abstractions upon which all simulation components must be built. Without this set of shared abstractions, independent simulation components cannot easily cooperate.

Simulacron implements discrete time MSM, meaning that the simulation is advanced by one or more "ticks". Ticks can be of uniform or varying length, and any integer number of milliseconds long.

Simulacron depends on two key types of entities for representing the world; one for locations and one for individuals. Locations are represented as "cells", which have no inherent properties other than a unique identifier. Individuals are "peeps", which are similarly minimal, knowing only their own identifier and which cell they are currently located in.

The data associated with each simulation process is attached to cells and peeps via "fields" which are effectively arbitrary packets of data associated by name. Although maintained by the framework, it does not possess or require any knowledge about the structure or contents of them. Finally, the framework also defines "state sets" which are named collections of fields, primarily used to modify the behaviour and state of individuals during the simulation. State sets can be used to represent the impact of policies on individuals, thus as policy changes, the associated state set is activated. There can only be one state set active at a time.

These five core abstractions have, thus far, proven a sufficient baseline for the construction of everything else.

In practice, the framework provides implementations of these constructs and other useful pieces of functionality. This includes data storage, serialisation, network communication, data input, and report output. In addition, there are a few code modules which, whilst not part of the framework itself (you can remove these modules and the framework will be unaffected), are still needed for various simulation components to function. These implement shared, but non-core, concepts such as abstract action representation and mortality.[2] The action system was added to provide a centralised mechanism for exposing operations on individuals, at the input data set level. Modules (see below) can expose trigger conditions to the simulation designer, allowing some predefined action to take place when said condition is met. For example, there could be a trigger in the event that an individual becomes infected by, or recovers from, a disease. This trigger might change the individual's current destination, or their active state set. On loading, modules register the actions which they implement with the action system, making them available to all other modules.

3.2   *The Simulacron Modules*

A number of simulation modules have been implemented atop the framework. Each module implements a logically distinct process which can operate on cells, peeps, or both. The data associated with each module is maintained by the framework in one or more fields defined by that module. Presently, there are modules which implement infection spread, terrorist/police interactions, scheduling of actions, path-finding, and basic rule-based motivation inspired by BDI. There are also a number of smaller modules implementing simple functionality such as generation of censuses, peep state, etc. The underlying module design principle we have adopted is to focus on *abstraction*, rather than *specialisation*. This leads to a far greater capacity for re-use and adaptation to scenarios beyond those originally envisaged. This generality can involve greater computational cost, but this has not proven to be problematic in practice. We also believe that the added design and implementation work is overwhelmingly worth the effort.

3.2.1   Scheduling

The scheduling module's purpose is to provide a way to define the movements and actions of peeps over time. The module provides two primary ways to control peeps: *cyclic schedules* and *event schedules*.

---

[2]Prior to the introduction of the mortality concept, peeps killed by the infection module would continue about their daily schedules. Whilst disturbing, this did not materially affect the results. Death is, it seems, widely useful but not always required.

In both cases, schedules contain one or more schedule entries. Each of these entries consists of a trigger condition and a standard action. The trigger condition is always a time stamp, although the interpretation of its value differs between the two kinds of schedules. In a cyclic schedule, each day, of which there may be many, interprets time stamps as a relative "time of day,"[3] whilst entries in event schedules are interpreted as absolute times.

Although there is nothing in the design of the simulation framework that precludes interactive manipulation of data, such a facility has not yet been implemented. As a result, complex sequences of events are specified in the data set prior to beginning the simulation using the scheduling functionality. Changes in peep behaviour are stored as state sets, with schedules triggering transitions between these states. Changing the active state set can, in turn, cause a different schedule to be used for subsequent ticks.

3.2.2 Infection

The infection module was designed to model a variant of the traditional SIR (Susceptible, Infected, Removed) infection model. The sequence of states supported by the model are: *Healthy Susceptible, Asymptomatic Latent, Asymptomatic Infectious, Heroic Infectious, Symptomatic Infectious*, followed by either *Immune* (which leads back to *Healthy Susceptible*) or *Dead*. "Heroic" is an original state which reflects the propensity for people to actively ignore or hide symptoms of an infection for various reasons. Consider the office worker who believes that "it's just a sniffle; nothing serious" or that they cannot afford to take leave from work.

In addition, the module features a conditional isolation feature which allows susceptible individuals to be sent to an isolation cell during specific period(s) of the day. An example is that sick children at a boarding school are not "detected" and isolated during the night, this will occur during the day when the teachers and other staff are awake. The module also has features for conditional infection, allowing the population to be divided into subgroups which can infect other, specific subgroups. This was designed to allow for basic multi-vector infections to be modelled.

Finally, support for multiple, concurrent infections has recently been added. Although it is not explicitly supported by the module, the presence or absence of one disease can influence an individual's susceptibility to another, through the state set mechanism.

3.2.3 Transport

The transport module provides a generalised routing and path finding mechanism, allowing individuals to navigate through a sequence of goal destinations. The navigation functionality is provided by a simple A* algorithm. As the basic cell

---

[3]That is, they discard the "date" portion of the time stamp.

has no concept of size or location, these are added through this module to allow such path finding to take place. This has largely replaced the previous movement mechanism, which was instantaneous teleportation.

3.2.4 Dispersion

The dispersion module was designed to provide a simple mechanism for non-deterministic movement of peeps. Over time, it has evolved into a more general mechanism for environmentally directed movement. The core functionality of this module is the random dispersion of peeps. Each cell with a dispersion field contains a set of one or more dispersion targets. On each tick, each peep in the cell is sent to a target randomly selected from this set.

The canonical example of this is a school playground. The playground is divided into four quadrants. Each quadrant is configured to disperse into the adjacent quadrants. With this arrangement, children peeps entering any part of the playground will randomly mill about until moved out of the playground by some other process.

In addition to the above, the dispersion module supports three additional mechanisms: masking, biasing and rate limiting. Masking allows the dispersal pattern to be varied depending on the specific peep being dispersed. For example, masking could be used to move wind peeps around an environment without causing person peeps to be affected. Biasing allows the relative probability of each target in the dispersion set to be adjusted. Rate limiting is used to control the amount of time between dispersal of successive peeps. This is often used to implement queues, barriers and other similar features.

3.2.5 Induced Peep Actions (IPA)

The IPA module allows for arbitrary actions (as defined by the action system) to be performed on peeps. One use of this module is to implement detours by redirecting peeps entering a certain cell to a new destination. In conjunction with the state set mechanism, this allows directed evacuation behaviours to be initiated and simulated.

3.2.6 Motivation

The motivation module provides a mechanism for peeps to interact with each other and their environment. The module allows peeps to have variables (called characteristics), which can be influenced by other peeps and the environment. Peeps may also have one or more if-then-else rules, which can execute actions. The values of a peep's characteristics determine which rules (and hence, actions) are executed. This has been used, for example, to model the spread of panic through a population. It does not provide for behavioural learning, being designed to be as simple as possible whilst still providing a useful level of fidelity. There is nothing to preclude the development of an additional module which supports learning and dynamic behaviour.

3.2.7 Terrorist, Police, Citizen (TPC)

This module enables the modelling of simple terrorist interdiction scenarios. To that end, it divides the population into three broad classes. Terrorists attempt to remain undetected by the police until their appointed attack time, whereupon they undertake whatever action has been specified (defaulting to killing everyone in the vicinity). Police attempt to detect and detain terrorists. Citizens play no active part in the simulation as far as this module is concerned.

3.3 Associated Tools

3.3.1 Data Set Template Processor (DSTP)

For the sake of flexibility, data set preparation is distinct from the simulation itself. Input data sets are plain-text XML files which completely define all entities and everything which is known about them; there are no emergent properties in the input. These files are the only input, aside from command-line parameters, which the simulation program accepts. The files are human-readable, allowing them to be written and modified by hand if necessary.

More commonly, however, they are constructed using templates. The template language is a set of XML pre-processing commands mixed with the raw data, interpreted by the DSTP program. These templates allow for the moderately straightforward creation of large populations. For example, instead of defining one thousand individuals, one can define a template that defines the statistical distribution of a population's properties, then sample 1,000 individuals from said population. The advantage of this method is that you do not lose the ability to manually construct unusual or special individuals. As with the generated data sets, the template files are plain text.

It is worth noting that the expansion of templates being independent of running the simulation also allows for random variation at two levels: a template can be re-instantiated with a different seed, producing a different, but similar, population. Alternatively, the same instantiated population can be re-simulated with a different simulation seed, leading to stochastic sampling of the behaviour spectrum of a fixed population.

Templates can also be extended with a simple built-in scripting language, or external programs. These allow for arbitrarily complex transformations to be performed without the need for additional, explicit steps in the process. As an example, one scenario required the calculation of spatial properties for the environment, a task handled by a simple external program written in the Python programming language and executed by DSTP.

DSTP allows templates to range from unmodified, static data sets (data sets with patterns abstracted) through to complex, multi-file, modularised templates with domain-specific representations and command-line replaceable definitions.

3.3.2 Refinery

This is used to estimate input parameters given known statistical measures. This is especially important for epidemiological purposes where suitable parameters, in the form we require, may not be available.

Refinery works by instantiating a population with parameters sampled from a pre-determined, broad distribution. Each instantiated population is then simulated a number of times. Simulated annealing is then used to refine the parameter ranges which are subjected to re-iteration. This process is repeated until convergence occurs, or until reasonable time or iteration constraints are exceeded.

Currently, Refinery can only be used to estimate a fixed set of parameters related to epidemiological modelling, but the principle is readily extensible to other parameters.

3.4 Simulacron in Practice

3.4.1 Royal Naval School Influenza

[6, 7, 12] The first case we studied was the influenza outbreak at the Royal Naval School in Greenwich, England during 1920. This was chosen as there was very detailed information on the outbreak and the conditions in which it occurred. The school itself had roughly one thousand students attending it, almost all of whom lived on-campus and rarely left the school grounds, making it an ideal, real-world closed community.

The scenario employed three of the simulation modules: the scheduling module which handled each individual's daily routine, the dispersion module which was used to model playground behaviour, and the infection module which codified the spread of the infectious disease. This scenario produced close agreement with historical data, even prior to the introduction of Refinery.

3.4.2 Australian Community Terrorism

[12] We have also applied Simulacron to the modelling of terrorist behaviours. There were two primary motivations for this work: to demonstrate the applicability of the modelling technique to problems outside of epidemiology, and determine whether the design of the simulator was sufficiently flexible to allow the existing framework and simulation modules to be adapted to a different purpose.

This involved the modelling of a statistically average Australian community which was subsequently used as the "background population" for the terrorism scenarios. The properties of both the individuals and the family units were chosen to match statistical data from the Australian Bureau of Statistics' 2006 census for New South Wales. Individuals moved about the community according to individually constructed schedules which emulated six groups of people: full-time employed, part-time employed, unemployed/home workers, primary school students, secondary school students, and infants. The distributions of individuals in family groups, residences and workplaces were also statistically matched.

Two scenarios were constructed atop this background community. The first involved the release of an infectious biological agent at a specific time and place within the community, with the spread of this agent measured. The infection module was used for this scenario, the biological weapon modelled as a stationary individual who was scheduled to become extremely infectious at the given attack time for a short duration. Our results suggest that location of release is a significant factor in determining the number of subsequent infections. This is hardly surprising given that different locations will have greatly varying traffic levels and patterns. Such simulations could be of interest for the purposes of risk assessment, provisioning resources to cope with an attack, and the effectiveness of potential interventions.

The second scenario examined the effectiveness of a simple interdiction strategy against terrorist activity. The goal of the terrorist in this scenario was to avoid detection until arriving at a club where they would detonate a bomb, killing all individuals in the building. Prior to the attack, the terrorist would engage in "normal" activities, consistent with others in the community. This was intended to mimic an individual attempting to avoid drawing attention to themselves. Police officers would move about the community on pre-planned patrol routes. The TPC module was initially developed for this scenario.

3.4.3 Amoy Gardens SARS

[9, 11] This simulation modelled the spread of severe acute respiratory syndrome (SARS) through Amoy Gardens in Hong Kong during March and April of 2003. This was chosen because it allowed us to study the effect that enforced isolation has on the spread of an infection, and it served as an opportunity to test and demonstrate our ability to model multi-vector infections. To that end, a synthetic community representing the residents of Block E was constructed, using demographic and work characteristics for a typical Hong Kong community. Although work on the scenario has not yet been completed, preliminary results demonstrate that the aerosol dispersal of infected droplets does play a significant role in such outbreaks.

It is worthy of note that both the humans and aerosol particles are modelled using peeps. The significantly different roles and behaviours of these two are managed via the masking mechanisms in the infection and dispersion modules.

These simulations were run on a relatively low-spec consumer laptop with one hardware thread. They involved approximately 15,000 individuals, who were simulated with the infection, scheduling and dispersion modules.

3.4.4 Sydney Central Station Terrorism

Our most recent completed work was in modelling a hypothetical bomb blast at Central Station in Sydney, Australia [8]. The purpose of this scenario was to perform a more thorough and involved test of the motivation module, provide an environment for an initial implementation of a more realistic path-finding movement module, and to demonstrate and assess our tools' capacity for modelling a realistic transportation hub.

This scenario involved the modelling of the upper level of Central Station. A synthetic commuter population of approximately half of peak-hour traffic was also created and simulated. Normal commuter behaviour was simulated for one hour before the introduction of an assailant peep. The assailant moved through the station to a pre-determined location before detonating an explosive device. The explosion killed or injured everyone in the immediate vicinity and was normatively equivalent to a low-yield backpack IED. The simulation continued until the two hour mark in order to observe how the commuters responded to this event.

The motivation module allowed the modelling of post-blast commuter behaviours consistent with those observed in similar, real-life events.

The explosion and subsequent effects of the bomb were modelled with the infection module by representing it as a high-lethality infection with no latent, asymptomatic or symptomatic periods, and no cross-infection.

Although this scenario was a proof-of-concept, it demonstrates a realistic capacity for our approach and tools to model events of this nature. With a modular framework and generalised components, a reasonable facsimile of a real-world location was simulated with a meaningful number of individuals, all on a single desktop computer.

These simulations were run utilising a single core of a stock Intel® Core™ i7 860 CPU, taking an average of 50 min to simulate 2 h with a temporal granularity of 5 s. These scenarios involved 10,000 commuters, simulated with a combination of infection, scheduling, dispersion, induced peep actions, path finding and motivation. As a point of comparison, a similar agent-based simulation involving evacuation of individuals from the International terminal of the Los Angeles International Airport involved just 200 representative individuals [18]. Although we do not have information regarding the computational resources required by this, the limited number of individuals modelled strongly suggests that they were not inconsiderable.

Conclusion

We believe that the success of Simulacron across multiple simulation domains demonstrates the value of the strategy we have adopted. It provides a light-weight mechanism for the implementation of heavy-weight simulations, through a focus on generality and modularisation, and enables reusability and composability.

The goals of the research team drive development from two directions. One is the provision of increasing flexibility and scope of modules, the other is the creation of more complex simulations. The ability to address both of these within a single team reflects the advantages and the challenges of working in a multi-disciplinary environment. The challenge of the project is to pursue the first without needlessly complicating the construction of simulations, and the second without creating a swathe of over-specific modules. To date, we believe these seemingly contradictory goals have been achieved, and we see no strong evidence that this will not continue to be the case.

In the near future, we plan to conduct an extensive re-write of the Simulacron framework, intended to further improve its performance and to make it more accessible to other researchers. This development will allow larger, more complex simulations across a broader range of problem domains. This will also facilitate further collaboration with outside agencies, increasing the breadth of utility of the framework. Ideally, this would include such agencies actively developing modules atop the framework.

Our design methodology thus far has been an organic one, meaning that we do not have any fixed direction for the future development or expansion of our simulation approach. As most of the enhancements made to-date (such as the introduction of state sets) were prompted by needs arising from the development of the software, it is hard to predict the details of future changes. However, the guiding principle of "keep it simple; if you can't, keep it general" is unlikely to change.

References

1. Balci, O., Arthur, J.D., Ormsby, W.F.: Achieving reusability and composability with a simulation conceptual model. J. Simul. **5**(3), 157–165 (2011). http://www.palgrave-journals.com/doifinder/10.1057/jos.2011.7
2. Birkin, M., Wu, B.: A Review of microsimulation and hybrid agent-based approaches. In: Heppenstall, A.J., Crooks, A.T., See, L.M., Batty, M. (eds.) Agent-Based Models of Geographical Systems, pp. 51–68. Springer, Netherlands (2012). http://dx.doi.org/10.1007/978-90-481-8927-4_3
3. Brouwers, L., Boman, M., Camitz, M., Mäkilä, K., Tegnell, A.: Micro-simulation of a smallpox outbreak using official register data. Euro Surveill. **15**(35), 1–8 (2010). http://www.ncbi.nlm.nih.gov/pubmed/20822732

4. Chen, X.: Microsimulation of hurricane evacuation strategies of galveston island. Prof. Geogr. **60**(2), 160–173 (2008). doi:10.1080/00330120701873645. http://www.tandfonline.com/doi/abs/10.1080/00330120701873645
5. Crooks, A.T., Heppenstall, A.J.: Introduction to agent-based modelling. In: Heppenstall, A.J., Crooks, A.T., See, L.M., Batty, M. (eds.) Agent-Based Models of Geographical Systems, pp. 85–105. Springer, Netherlands (2012). doi:10.1007/978-90-481-8927-4_5. http://dx.doi.org/10.1007/978-90-481-8927-4_5
6. Dudley, S.F.: The Spread of "Droplet infection" in Semi-Isolated Communities. H. M. Stationery Office (1926)
7. Green, A.R., Zhang, Y., Piper, I.C., Keep, D.R.: Application of microsimulation to disease transmission and control. In: Proceedings of the 2011 MODSIM International Congress on Modelling and Simulation, pp. 919–925 (2011). http://www.mssanz.org.au/modsim2011/B2/green.pdf
8. Keep, D., Piper, I.C., Green, A.R., Bunder, R.: Modelling of behaviours in response to terrorist activity. In: Chan, F., Marinova, D., Anderssen, R.S. (eds.) 19th International Congress on Modelling and Simulation, pp. 475–481. Modelling and Simulation Society, Perth (2011). http://www.mssanz.org.au/modsim2011/A6/keep.pdf
9. Keep, D.R.: Generalised Microsimulation. Ph.D. thesis, University of Wollongong (2012)
10. Orcutt, G.H.: A New Type of Socio-Economic System. Rev. Econ. Stat. **39**(2), 116 (1957). doi:10.2307/1928528. http://www.jstor.org/stable/1928528?origin=crossref
11. Outbreak of Severe Acute Respiratory Syndrome (SARS) at Amoy Gardens, Kowloon Bay, Hong Kong; Main Findings of the Investigation. Tech. rep., Health Department of the Hong Kong Special Administrative Region of the People's Republic of China Government Information Centre (2003)
12. Piper, I., Keep, D., Green, T., Zhang, I.: Application of microsimulation to the modelling of epidemics and terrorist attacks. Informatica **34**, 141–150 (2010). http://www.freepatentsonline.com/article/Informatica/232715912.html
13. Sauerbier, T.: UMDBS - A new tool for dynamic microsimulation. J. Artif. Soc. Soc. Simul. **5**(2) (2002). http://jasss.soc.surrey.ac.uk/5/2/5.html
14. Smith, L., Beckman, R., Baggerly, K., Anson, D., Williams, M.: TRANSIMS: Transportation analysis and simulation system. Tech. Rep. 26295, Los Alamos National Lab., NM (United States) (1995). http://www.osti.gov/energycitations/product.biblio.jsp?osti_id=88648
15. Spielauer, M.: What is social science microsimulation? Soc. Sci. Comput. Rev. **29**(1), 9–20 (2010). doi:10.1177/0894439310370085. http://ssc.sagepub.com/cgi/doi/10.1177/0894439310370085
16. Statistics Canada: Modgen (2011). http://www.statcan.gc.ca/microsimulation/modgen/modgen-eng.htm
17. Stroud, P., Del Valle, S., Sydoriak, S., Riese, J., Mniszewski, S.: Spatial dynamics of pandemic influenza in a massive artificial society. J. Artif. Soc. Soc. Simul. **10**(4), 9 (2007). http://jasss.soc.surrey.ac.uk/10/4/9.html
18. Tsai, J., Fridman, N., Bowring, E., Brown, M.: ESCAPES-Evacuation Simulation with Children, Authorities, Parents, Emotions, and Social comparison. In: Tumer, Y., Sonenberg, S. (eds.) Proceedings of the 10th International Conference on Autonomous Agents and Multiagent Systems. Innovative Applications Track (AAMAS 2011), pp. 457–464 (2011). http://www.aamas-conference.org/Proceedings/aamas2011/papers/D3_G57.pdf
19. Wolshon, B., Resiliency, T., Lefate, J., Naghawi, H., Montz, T., Dixit, V.: Application of TRANSIMS for the multimodal microscale simulation of the new orleans emergency evacuation plan. Tech. rep., Louisiana State University (2009). http://www.ntis.gov/search/product.aspx?ABBR=PB2011107618
20. Wu, B.M., Birkin, M.H., Rees, P.H.: A dynamic msm with agent elements for spatial demographic forecasting. Soc. Sci. Comput. Rev. **29**(1), 145–160 (2010). doi:10.1177/0894439310370113. http://ssc.sagepub.com/cgi/doi/10.1177/0894439310370113

From Global Polarization to Local Social Mechanisms: A Study Based on ABM and Empirical Data Analysis

Zhenpeng Li and Xijin Tang

Abstract Many studies show that social influences promote group polarization. In this paper we investigate the microscopic social mechanisms through agents based modeling and empirical data analysis. Both suggest that three types of social influences give rise to the emergence of macroscopic polarization, and the polarization pattern is closely link with local structure balance.

Keywords Agents based modeling • Empirical data analysis • Macroscopic polarization • Social influences • Structure balance

1 Introduction

Global polarization are widely observed in human society. Examples about the group behaviors patterns include culture, social norms, political election and online public hotspots debates. From decision-making perspective, the global pattern of collective dynamics is rooted from individuals' micro-level decision making processes, and social influence as one of the important social psychological factors, play dominant role in the processes. Generally to say, from the social influence point of view, three types of impact run through the whole processes of group decision-making especially in voting. One is positive influence, which may accelerate intragroup opinion convergence. The second one is negative social impact which may block the formation of consensus among different intergroup. The third type of special individuals' attitudes, in which the individuals might not belong to any labeled subgroup, members in the group have no common social identity, no firm stand about some opinions and are in a state of neither fish nor fowl [1–3].

Z. Li
Dali University, Dali, China
e-mail: lizhenpneg@amss.ac.cn

X. Tang (✉)
Chinese Academy of Sciences, Beijing, China
e-mail: xijintang@amss.ac.cn

© Springer Japan 2015
Q. Bai et al. (eds.), *Smart Modeling and Simulation for Complex Systems*, Studies in Computational Intelligence 564, DOI 10.1007/978-4-431-55209-3_3

From a bottom-up point of view, in modeling social processes, individuals' locally cumulative interacting behaviors would evolve into different global patterns. The emergence of global features come from the local interconnecting mechanism, e.g., short average path and high clustering coefficients contribute to small world mechanism [4]. However, it is difficult to infer how the global patterns are generated from the aggregation of certain simple local social processes. Then, Agent-Based Model (ABM) simulation techniques are widely applied. In this study, two different agents based models are used to investigate how social influences lead to group polarization, and the intrinsic link between local influence structure balance and the emergence of global bi-polarization pattern. Furthermore, we use on-line empirical data analysis to verify the simulation conclusion.

The rest of the paper is organized as follows: in Sect. 2 we detect the negative social influence mechanism for group bi-polarization by Voter Model (VM), Sect. 3 examines the relationship between social influence and bi-polarization through Hopfield Attractor Model (HAM). We especially focus on the connection between global polarization pattern and dyad/triad balance at the micro-level. In Sect. 4 we conduct empirical data analysis to show that our models make sense to explain the real world phenomenon, even if we do not expect these models could explain everything about complex group dynamics in our social world. Section 4.2 is our concluding remarks.

2 Voter Model

In this section, we first introduce simple Voter Model and assign three types of social influences on the social ties, then simulate the group opinions dynamics on the topology of lattice with periodic boundary.

2.1 *Voter Model and Its Revised Version*

The Voter Model is a simple mathematical model of opinion formation in which voters are regarded as the nodes of a network, each voter (individual) has an opinion (in the simplest case, -1 or $+1$). For a randomly chosen voter, his/her opinion has the chance to be altered by the opinions of the neighbors. It is often used to see how ordered states can appear in systems originally in a state of non-equilibrium. The model has several applications in a variety of disciplines including chemistry (reactions between different chemicals), physics (interactions between particles) and social systems (interactions between agents) [5–7]. The model formula is Eq. (1)

$$\omega(-\sigma_i|\sigma_i) = \frac{\beta}{2}\left(1 - \frac{1}{k}\sum_{j \in n(i)} \sigma_j \sigma_i\right), \tag{1}$$

where β is a constant adjustable parameter, $\sigma_i = +1$ or $-1 (i = 1, \ldots, N)$, N is the group size, voter $i's$ opinion is "$+1''$ means "for", "-1" means "against", $n(i)$ denotes for voter $i's$ neighbors, k is the size of neighbors. The left term of Eq. (1) is the probability that individual i might change its opinion from ± 1 to ∓ 1. There is a large number of literatures on the relationship between voter model and different topologies, such as scale-free, small world and lattices. Especially in lattices, the voter model presents simple non-equilibrium dynamics with nontrivial behavior, hence this model has been extensively studied, most notably by the mathematicians and condensed matter physicists, from various aspects.

In order to further understand the relationship between non-positive social influence and group opinions polarization quantitatively, we adopt three kinds of social influence mechanism on the classic Voter Model, and have classic Voter Model revised form [see Eq. (2)] in the topology of lattices. Individuals (voters) are impacted by others and also influence others, as conditioned by valence of the social identity tie I_{ij} :

(A) "+" implies attraction (homophily, similarity) and imitation,
(B) "−" stands for xenophobia (hetereophily instead of homophily) and differentiation (instead of imitation),
(C) "0" denotes for unsocial attitudes.

$$\omega(-\sigma_i|\sigma_i) = \frac{\beta}{2}(1 - \frac{1}{k} \sum_{j \in n(i)} I_{ij}\sigma_j\sigma_i), \tag{2}$$

where $I_{ij} = \{+1, -1, 0\}$.

2.2 Simulation and Discussion

In our simulation we set $\beta = 1$, group size $N = 10,000$, the neighbor size $k = 8$, the topology is 100×100 periodic boundaries lattices, $T = 10^3$.

Figure 1 shows that the classic voter model final "+1" percentage fluctuation against time step t. It is obvious that for each run (we run 7 times in the simulation) in the end, group opinion prone to reach one polarization steady state even though all runs begin with 50 % "+1" initialization.

Figure 2 shows the revised Voter Model with three kinds of social influence, final "+1" percentage tend to 50 % (or the ratio of against versus support approximate to 1 : 1), even though in the case with different initial percentage "+1".

The results show that the consensus could not occur with the non-positive social influence as does on regular two-dimensional lattices, instead the system settles in a stationary state with coexisting opinions two-polarization. The original voter model emphasized the global stability of social homogeneity, where convergence to one leading polarization is almost irresistible in closely interacting populations. However, the voter model with some influence ties to be non-positive suggests that social homogeneous stable state is highly brittle.

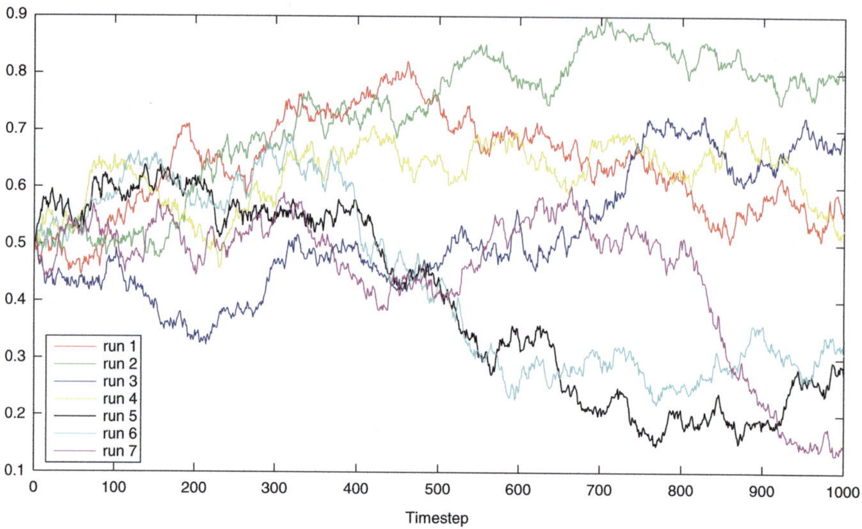

Fig. 1 Classic voter model one polarization dominant state with time evolution (50 % for or "+1" initialization)

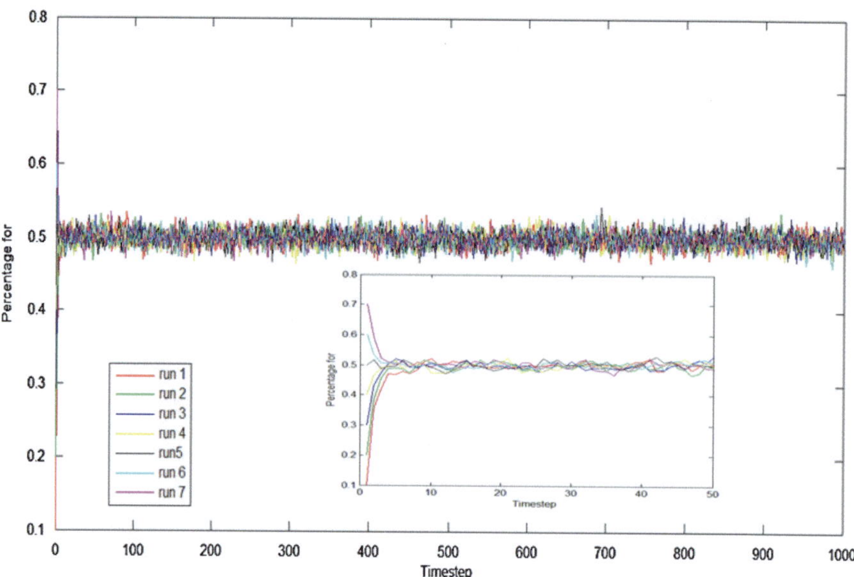

Fig. 2 Revised voter model two polarization steady state against time t (the inset figure show the convergent trend for the seven times different initialization of percentage for)

3 Hopfield Attractor Model with Dyadic Influence Ties

3.1 The Model

Macy et al. (2003) presented a Hopfield Attractor Model (HAM) to describe group polarization problems, when considering individual decision making dimensions, social ties signs, strengths and culture dissemination theories, etc. The mechanism of their modeling is as following,

$$P_{is} = \frac{\sum_j I_{ij} s_j}{N}, j \neq i, \tag{3}$$

for each individual i, the cumulative social pressure for her/him to choose s_i is denoted as Eq. (3), where $s_i = \pm 1$ represent binary voting opinions, N is group size, I_{ij} is the social influence that individual j imposed to i, matrix $(I_{ij})_{N \times N}$ is named social influence matrix. According to HAM, with the motivation of investigating the relationship between non-positive social influence and group opinions polarization, instead assigning continuous values between -1 and $+1$ to I_{ij} as in [8], we assign three discrete values$-1, +1, 0$ to I_{ij} to indicate the three types of dyadic social influence defined in [1], and have Eq. (4) logistic form social pressure,

$$\tau_{is} = \frac{v_s}{1 + e^{-KP_{is}}} + (1 - v_s)X_i, \tag{4}$$

where $X_i(-0.5 \leq X_i \leq 0.5)$ denotes for the external intervention, and $v_s(0 \leq v_s \leq 1)$ is used to trade off the internal and external group influence for individual i opinion. K is the size of opinions dimension. Given a randomly selected threshold π, if $\tau_{is} \geq \pi$, then individual i chooses "$+1$", else i chooses "-1". Equation (5) describes the updating influence processes of individual $j(j \neq i)$ to i.

$$I_{ij}(t + 1) = I_{ij}(t)(1 - \lambda) + \frac{\lambda}{K} \sum_{k=1}^{K} s_{jk}(t)s_{ik}(t), j \neq i. \tag{5}$$

Where t is the time step, λ is an adjustable parameter between 0 and 1.

Based HAM LI and TANG consider the situation of discrete social ties, and endow specific social influence connotation to the abstractive social ties. In addition, they extended the dyadic influence structure to triadic scenarios. They also discovered the intrinsic connection between local dyadic and triadic structures balance and global bi-polarization pattern. The other interesting finding is that "0" the neutral relationship tend to vanish with step t update, when bi-polarization come into being "0" completely convert to negative or positive relation [1,2]. Next we will present simulation result on [1].

3.2   Simulation Results Analysis

We take the test by setting $N = 100$, $K = 5$, π belong to $[0,1]$ uniform distribution and $\lambda = 0.5$. We run 100 times for average. Figure 3a shows the group initial random opinions states when each agent i face K-dimension decision making (before self-organizing polarization). Figure 3b illustrates the group bi-polarization state under the condition of no imposing external influence ($v_s = 1$) and with three types of influence. We can observe that two patterns appear after group polarization, i.e., one pattern is $(+1, +1, 1, 1, 1)$, i.e. (white, white, black, black, black) (marked by v_1), the other is $(1, 1, +1, +1, +1)$, i.e., (black, black, white, white, white) (represented by v_2). The ratio of the 2-pattern size is approximate to 1:1.

The relationship between exogenous intervention parameter v_s to group polarization is as shown in Fig. 4.

We can see that when $v_s = 1$ (no external intervention to the group interaction processes), the ratio of is $v_1/(v_1 + v_2)$ approximate to 0.5. However, the fifty to fifty well matched equilibrium will be destroyed with a little cut off v_s. In other words, external intervention will lead to majority dominant pattern. In particular, when $v_s = 0.5$ group opinion is evenly affected by external and internal factors, we observe the group consensus appears, i.e., $v_1/(v_1 + v_2)$ is approximate to 1, the pattern v_2 nearly disappears. It is clearly suggested that, under the condition of imposing external intervention, the group reaches majority or consensus pattern. With no exogenous impact, the group evolves into bi-polarization state in the end.

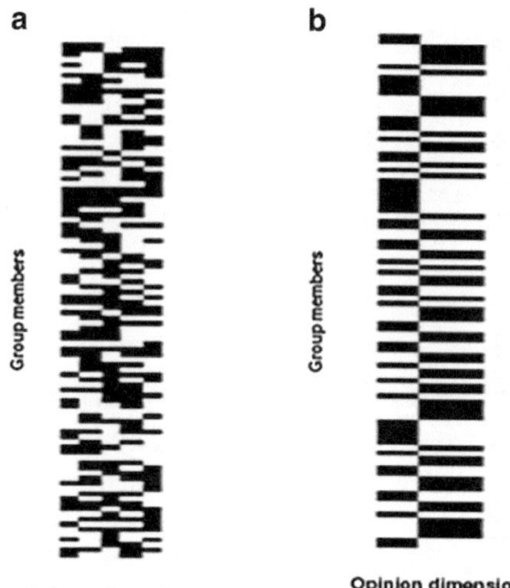

Fig. 3 Group opinion before and after polarization under the condition of imposing three types of social influence. (We generate $N \times K$ matrix where N denotes group size and K denotes the numbers of options, $N = 100$, $K = 5$)

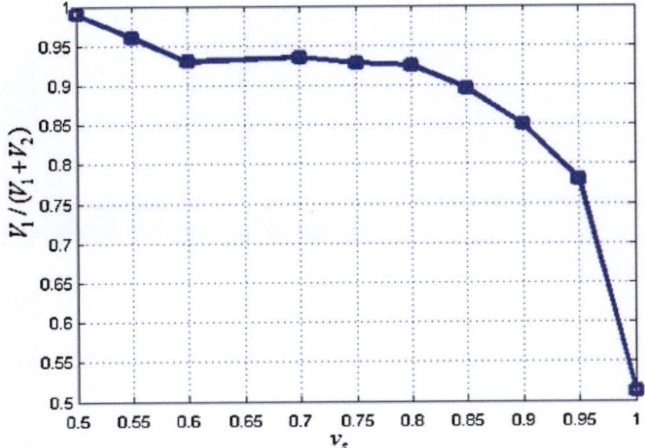

Fig. 4 Exogenous intervention impact on group polarization (In Eq. (4) we adjust vs from 1 to 0.5 i.e., increase the out-group intervention from 0 to 0.5, to observe the exogenous effect on group opinions polarization)

Furthermore, we investigate the triadic relation motifs distribution before and after bi-polarization by using R package sna [9]. We find that the overwhelming structure balance motifs[1] emerge concurrently with polarization process.

Figure 5 presents the dynamic variation of triadic distributions. At the very beginning $t = 0$, the upper plot in Fig. 5 shows the initial local triads distribution according to randomly generated social influence matrix. We can observe that all 16 types of triads exist in the initially. With the social influence matrix updating, at step $t = 19$ some triad motifs disappear, e.g., Code 003, Code 012 and Code 102, while Code 300, Code 210 become dominant (see middle plot in Fig. 5). At step $t = 29$, other triad motifs disappear except balanced triad motif Code 300 (see bottom plot in Fig. 5). Both simulation results based on Voter Model and Hopfield Attractor Model suggest that non-positive social influence promotes group bi-polarization pattern. In addition, HAM modeling result concludes that opinions polarization in a group is coexisted with local level structure balance, which reveals some interesting internal connection between global collective pattern and local social structure stability. Next we try to use ABM simulation conclusions to explain some real world collective patterns, although we do not expect that the simple computing models could cover all aspects of our complex group behaviors.

---

[1] Holland and Leinhardt (1970) addressed that classic balance theory offers a set of simple local rules for relational change and classified local triadic motifs into 16 types, according to mutual reciprocity, asymmetry relation, non-relationship between pairs (or dyadic relations), where Code 300 triad relation corresponding to structure balance under the condition of the triad product signs satisfies "+". More detail about structure balance please refer to [10–12].

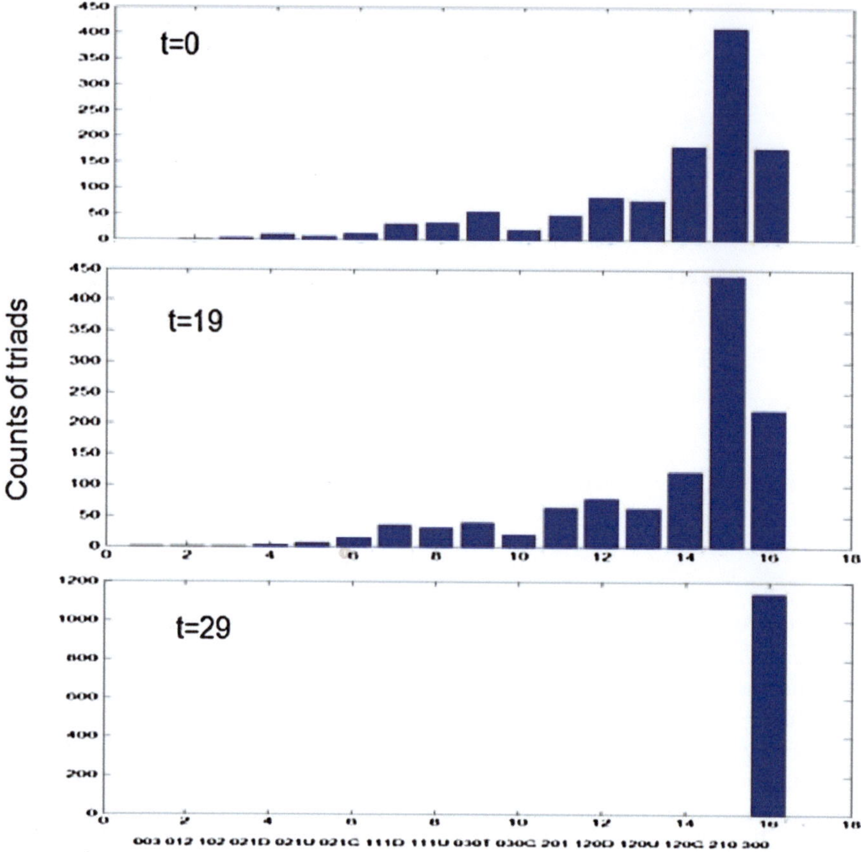

Fig. 5 Sixteen types triads dynamic distribution versus updating step *t*

4   The Evidences from Empirical Data Analysis

In this section, we present two empirical illustrations, one show that no-positive influence as an important factor that promote political opinions polarization over social media, the other suggest that on-line social network demonstrate the local structure stable characteristic.

4.1   Political Polarization on Twitter

Social media play important roles in shaping political discourse in the U.S. and around the world. Some empirical evidence suggests that politically active Internet users tend to organize into insular, homogenous communities segregated along

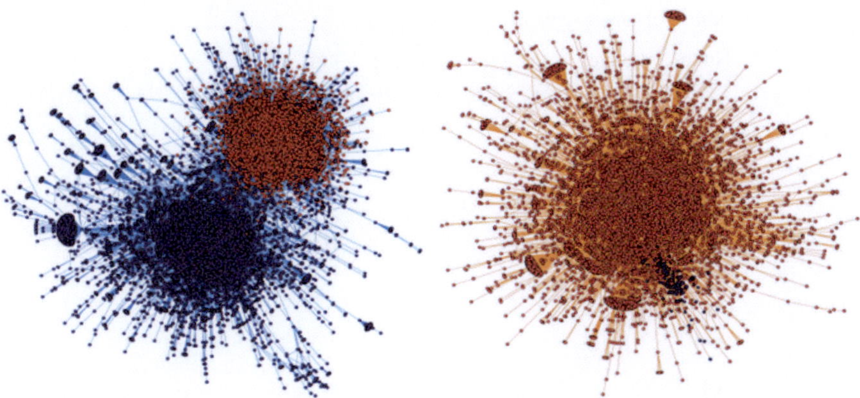

Fig. 6 The political retweet (*left*) and mention (*right*) networks. Node colors reflect cluster assignments. Community structure is evident in the retweet network, but less so in the mention network. In the retweet network, the *red cluster* A is made of 93 % right-leaning users, while the *blue cluster* B is made of 80 % left-leaning users [15]

partisan lines. For example, Adamic and Glance (2005) demonstrated that political blogs preferentially link to other blogs of the same political ideology [13]. That is also supported by the work of Hargittai et al. (2007) [14], that Internet users who share the same political point of view tend to behave similarly, and choose to read blogs that share their political beliefs. More insightful evidence is from Conover et al. (2010), who examine networks of political communication on the Twitter during the six weeks prior to the 2010 U.S. midterm elections. This study shows that the retweet network exhibits a highly modular structure, segregating users into two homogenous communities corresponding to the political left and right, demonstrate obvious bi-polarization characteristic. Most surprising contrast is that the mention network does not exhibit the polarization pattern, resulting in users being exposed to individuals and information they would not have been likely to choose in advance [15]. Figure 6 displays this kind of group opinions dynamic pattern.

Those findings suggest that the possibility that in-group/out-group differentiation and rejection antagonism, or the non-positive social influence between intragroup members and intergroup members are the emergent cause of social network self-organization, and lead to the bi-polarization global dynamic pattern.

4.2 Local Structure Balance on Epinions Trust Network

We investigate the local structure balance among mixed relationships, i.e., positive, negative, and neutral by using Epinions dataset. Epinions product review Web site is a trust network, where users can indicate their trust, distrust, or no comments of the reviews of others. In addition, people can give both positive and negative ratings not

Table 1 Dataset statistics

Nodes	Edges	+Edges	−Edges	+Δ	−Δ
131,828	841,372	717,129	123,670	33,682,027	2,994,773

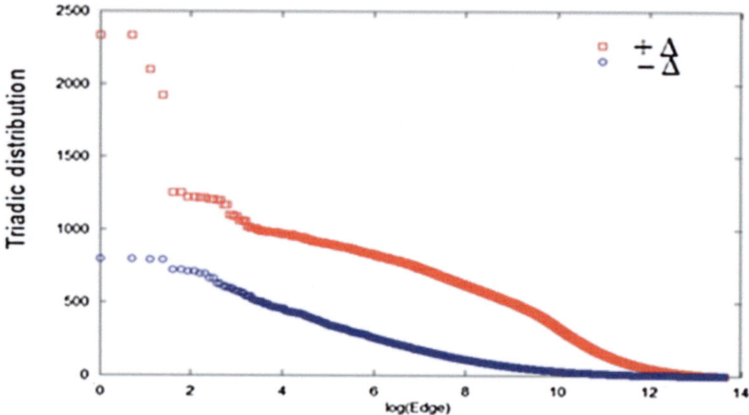

Fig. 7 Balanced triads and unbalanced triads distribution

only to items but also to other raters. In our study, we analyze the Epinions dataset with 131,828 nodes and 841,372 edges. The basic statistics about Epinions data set is listed as in Table 1.

In Table 1, "+" stands for the triadic relationship that satisfy structure balance, "−" denotes for the unbalance triadic relation. "+Edges" stand for the trust relationship, "−Edges" denotes for the opposite. Figure 7 shows that most of the edges are embedded into smaller size community (subgroup), and only a few edges are embedded into high density modular structure. However, it is worth noting that in both scenarios balanced triads distribution is higher than that of unbalanced counterparts.

We also analyze the 16 types triads distribution for three cases (trust or positive "+", untrust or non-positive "−", mixed or including three scenarios no comments relationships by "0", and "+", "−"), correspondingly. Our findings is that the size of local triadic balance structure such as 16-Code300, 3-Code102, 11-Code201[11] in the network constructed by trust edges (or positive "+" relationship) and mixed relationship (including "+", "0", and "−" links) is more than that of in the untrust (or negative "−" relationship) network. These observations fit well into Heider's original notion of structural balance, which in its strongest form postulates that when considering the relationships between three people, either only one or all three of the relations should be positive, see Fig. 8.

In summary, we find that the data analyzing results is closely connected to the theory of classic social psychology, structure balance which tends to capture certain common patterns of group interaction.

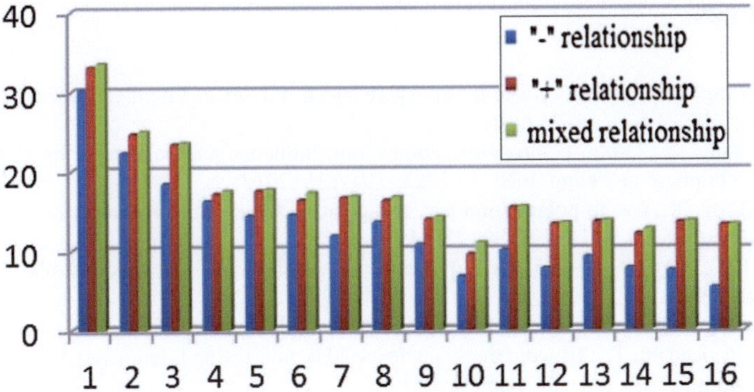

Fig. 8 Sixteen types triads statistic distribution of the three networks in Epinions (y-axis is plotted in log-scale)

Conclusion

In this study, we investigate the non-positive social impact on group polarization based on HAM and VM. By simulation, we show that bi-polarization pattern tends to emerge with no imposing external intervention, and consensus may occur among group members if the non-positive influence is neglected. With both dyadic and triadic influence considered, in HAM simulation we observe that dyadic and triadic influence balance among agents has inherent relation with global bi-polarization pattern. It is similar to the macro-micro linkage: sentiment relations among agents (localized as triad and dyad) lead to the collective balance of the group. Two empirical evidences are consistent with our modeling results, polarization on twitter shows that no-positive influence as important factor that promote voters' opinions polarization, Epinions trust network suggests that on-line signed social network demonstrate the local structure stable characteristic. Why are some communities, organizations, and social group characterized by cooperation, trust, and solidarity, i.e. consensus pattern, whereas others are fraught with corruption, discord, and fear, i.e. unconformity or polarization pattern? We argue that from the bottom-up view, the answer centers on global pattern that has its micro-level social mechanisms, by these principles, decision making is distributed and global order self-organizes out of multiplex local interactions among autonomous interdependent actors. Evidences in China social media are our next endeavors.

Acknowledgements This research was supported by National Basic Research Program of China under Grant No. 2010CB731405, National Natural Science Foundation of China under Grant No.71171187.

References

1. Li, Z., Tang, X.: Polarization and non-positive social influence: a hopfield model of emergent structure. IJKSS **3**(3), 15–25 (2012)
2. Li, Z., Tang, X.: Group polarization: connecting, influence and balance, a simulation study based on hopfield modeling. PRICAI **2012**, 710–721 (2012)
3. Li, Z., Tang, X.: Group polarization and non-positive social influence: a revised voter model study. Brain Inform. **2011**, 295–303 (2011)
4. Watts, D.J., Strogatz, S.: Collective dynamics of "small-world" networks. Nature **393**(6684), 440–442 (1998)
5. Clifford, P., Sudbury, A.: A model for spatial conflict. Biometrika **60**(3), 581–588 (1973)
6. Holley, R., Liggett, T., : Ergodic theorems for weakly interacting infinite systems and the voter model. Ann. Probab. **3**(4), 643–663 (1975)
7. Castellano, C., Vilone, D., Vespignani, A.: Incomplete Ordering of the voter model on Small-World Networks. EuroPhys. Lett. **63**(1), 153–158 (2003)
8. Macy M,W., Kitts, J.A., Flache, A.: Polarization in dynamic networks a hopfield model of emergent structure. In: Breiger, R., Carley, K., Pattison, P. (eds.) Dynamic Social Network Modeling and Analysis: Workshop Summary and Papers, pp. 162–173. The National Academies Press, Washington (2003)
9. Butts, C.T.: Social Network Analysis with sna. J. Stat. Softw. **24**(6), 1–51(2008)
10. Heider, F.: Attitudes and cognitive organization. J. Psychol. **21**, 107–112 (1946)
11. Holland, P.W., Leinhardt, S.: A method for detecting structure in sociometric data. Am. J. Sociol. **70**, 492–513 (1970)
12. Cartwright, D., Harary, F.: Structural balance: a generalization of Heider's theory. Psychol. Rev. **63**, 277–292 (1956)
13. Adamic, L., Glance, N.: The political blogosphere and the 2004 U.S. election: Divided they blog. In: Proc. 3rd Intl. Workshop on Link Discovery (LinkKDD), pp. 36–43 (2005)
14. Hargittai, E., Gallo, J., Kane, M.: Cross ideological discussions among con-servative and liberal bloggers. Public Choice **134**(1), 67–86 (2007)
15. Conover, M.D., et al.: Political polarization on twitter. In: Proceedings of the 5th International Conference on Webblogs and Social Media, pp. 237–288, Barcelona (2011)

Decentralised Task Allocation Under Space, Time and Communication Constraints in Disaster Domains

Xing Su, Minjie Zhang, and Quan Bai

Abstract The coordination of dynamic task allocation based on available resources is a very challenging issue in disaster domains under time, space and communication constraints. In addition, it is also very hard or even impossible to achieve tasks allocation in a centralised manner with the global knowledge of such an environment. This paper presents a novel decentralised coordination approach for dynamic task allocation by considering space, time and communication constraints in a disaster domain, and workloads and priorities of different tasks. In this approach, a group formation mechanism is proposed to help agents with limited communication ranges to achieve efficient task allocation in a group through cooperation. The overall task allocation is achieved through distributed coordination in each dynamic group without a central control mechanism to reflect real life situations in a general disaster domain. The experiment results show that the proposed approach outperforms other decentralised approaches, in disaster domains under space, time and communication constrains.

Keywords Agent coordination • Disaster environments • Intelligent agents • Task allocation

1 Introduction

Nowadays, agent-based coordination for task allocation has been widely applied in many domains such as disaster rescue, space exploration and distributed computing, etc. [6, 12, 15]. The main objective of task allocation is to allocate the most suitable resources (usually agents) to different tasks in a rational way. In addition to the main

X. Su (✉) • M. Zhang
University of Wollongong, Wollongong, NSW, Australia
e-mail: xs702@uowmail.edu.au; minjie@uow.edu.au

Q. Bai
Auckland University of Technology, Auckland, New Zealand
e-mail: quan.bai@aut.ac.nz

© Springer Japan 2015
Q. Bai et al. (eds.), *Smart Modeling and Simulation for Complex Systems*, Studies in Computational Intelligence 564, DOI 10.1007/978-4-431-55209-3_4

objective, there are different new requirements and constraints, which are requested for the efficient coordination of task allocation in different application domains. This paper focuses on task allocation problems in disaster domains.

Normally, a disaster domain has the following major constraints and issues for task allocation. (1) **Space constraints**. In a disaster domain, tasks are discovered at different locations. Hence, agents need to move to the location of tasks, which will consume time. (2) **Time constraints**. In a disaster domain, the objectives of task allocation include locating and saving survivors in debris, extinguishing a fire before it spreads, etc. Under such circumstances, each task should have a deadline and the task should be finished before the deadline (i.e. when the survivor is alive or the building is still stand). (3) **Communication constraints**. Due to the destruction of local infrastructures and other conditions in a disaster domain, rescue agents have difficulties to communicate with each other. (4) **Task Urgency**. Different tasks have different degrees of urgencies. The most urgent tasks need to be finished before the deadlines, and the least urgent tasks can be disregarded during task allocation if resources are not sufficient. In many existing approaches [5, 11], the researchers emphasised on finishing tasks as many as possible before the deadline, but did not consider the degree of urgency of tasks. For example, supposing there are two collapsed buildings, one with ten survivors and the other with 10 tons of goods. When allocating agents for these two buildings, rescuing the ten survivors would take precedence over saving the 10 tons of goods if the available resources are not sufficient to do both. The degree of urgency of each task is obviously a key issue that should be considered.

To handle the above constraints and issues, various models, mechanisms and approaches have been proposed to assist coordinating task allocation in a disaster domain each having its own approach [2, 7, 8, 13]. A number of centralised approaches [5, 11] have been developed to coordinate task allocation in a diaster domain. The centralised approaches can ensure an optimal allocation solution if a central controller can have global knowledge of overall tasks and agents in the domain. However, in most disaster domains, it is hard for a central controller to have that kind of knowledge due to the dynamics of the agents and tasks, as well as time and communication constraints.

To overcome the limitations of centralised approaches, some decentralised approaches [1, 3] have been developed to deal with the coordination of task allocation in a diaster domain. One of the famous models is the F-max-sum algorithm proposed by Ramchurn et al. [10]. The F-max-sum algorithm employs message passing between agents to achieve task allocation. If communication quality and range are limited, it is hard for agents to get enough messages from each other to get optimal solution for task allocation so the F-max-sum algorithm cannot work well. In addition, the F-max-sum algorithm does not consider the different urgencies of tasks. Actually, in most of the current research, the communication constraints only refer to the number of messages passing between agents and few of them consider the communication range of agents.

In order to meet the challenges of task allocation in disaster domains, a novel decentralised coordination approach for dynamic task allocation is proposed in this paper. Our approach employs a token passing mechanism to assist neighbouring agents to communicate with each other under communication constraints and to form groups in a decentralised manner. The contributions of the proposed approach are that (1) our approach considers space, time and communication constraints to reflect the real situation in a disaster domain; (2) our approach considers the degree of urgency and the workload of each task as well as the dynamic nature of a disaster environment so as to meet the conditions of the environment; (3) in our approach, a novel group formation mechanism is developed to help agent to form groups despite communication limits; and (4) a comprehensive utility function is designed to help the coordinator to find the best task allocation for the group.

The rest of paper is organized as follows. The problem is formulated and definitions are given in Sect. 2. The basic procedure of our approach is introduced in Sect. 3. The experiment and analysis are given in Sect. 4. Some related work about task allocation is introduced and analyzed in Sect. 5. The paper is concluded in Sect. 5.

2 Problem Description and Definition

In general, agent-based task allocation models the problem as there are a set of tasks and a set of agents. A task set contains m tasks, which can be described as $\{T_1, T_2, T_3, \ldots, T_m\}$, where T_i represents the ith task and $1 \leq i \leq m$. An agent set contains n agents, which can be described as $\{A_1, A_2, A_3, \ldots, A_n\}$, where A_j represents the jth task and $1 \leq j \leq n$.

In our approach, we give the following definitions to describe the problem in details.

Definition 1. A *Task* (T_i) can be defined as a six-tuple $T_i = <TNo, DLine_i, WLoad_i, Loc_i, Emg_i, TSta_i >$, where TNo is the ID of task T_i, $DLine_i$ is the deadline of T_i (e.g. the time point until which the survivor remains alive or the building remains standing), where $Dline_i \in [0, \infty]$, $WLoad_i$ is the workload of T_i, which represents how many work units must be done in order for T_i to be completed, Loc_i is the location of T_i, Emg_i is the degree of urgency of T_i, where $Emg_i \in [1, 10]$, in which 1 and 10 represent the lowest and the highest degrees of urgencies, respectively, and $TSta_i$ is the status of T_i, as either "available", "working", "finished" or "expired".

Definition 2. An *Agent* (A_j) can be defined as a six-tuple $A_j = < ANo, Uti_j, Loc_j, MSp_j, Comm_j, ASta_j >$, where ANo is the ID of A_j, Uti_j is the working efficiency of A_j. Uti_j represents how many units of workload A_j can perform per time unit. Loc_j is the current location of A_j, MSp_j represents the moving speed of A_j, which represents how far A_j can move per time unit, $Comm_j$ is the communication of A_j, and $ASta_j$ is the status of A_j, which can be either "available" or "working".

The main objective of weighted task allocation is to find an arrangement of agents to tasks to maximize the sum of weighted workload of finished tasks, which is described as:

$$Objective = argmax \sum_{i=1}^{m} Finish(T_i) \times WLoad_i \times Emg_i, \tag{1}$$

where $Finish(T_i)$ is a Boolean-valued function, if T_i is finished, $Finish(T_i) = 1$, otherwise $Finish(T_i) = 0$.

Apart from the main objective, the cost of agents for finishing tasks should be minimized, which includes (1) minimizing the traveling time and distance of agents to finish tasks and (2) minimizing the working time of agents to finish tasks.

As mentioned before, we employed a token passing mechanism to help agent efficiently pass information about tasks and agents. There are two types of tokens in our approach: task tokens ($TToken_k$) and agent tokens ($AToken_j$), both of which are created by agents. The two kinds of tokens are defined by Definition 3 and Definition 4, respectively.

Definition 3. A *Task Token* ($TToken_i$) can be defined as a six-tuple, $TToken_i =< ANo, TNo, DLine_i, WLoad_i, Loc_i, Emg_i >$, where ANo is the ID of the agent, which creates the $TToken_i$, TNo is the ID of the task represented by $TToken_i$, $DLine_i$ is the deadline of the task represented by $TToken_i$, $WLoad_i$ is the workload of the task represented by $TToken_i$, Loc_i is the location of the task represented by $TToken_i$ and Emg_i is the degree of the urgency of the task represented by $TToken_i$.

Definition 4. An *Agent Token* ($AToken_j$) can be defined as a four-tuple $AToken_j =< ANo, Uti_j, Loc_j, MSp_j >$, where ANo is the ID of the agent represented by $AToken_j$, Uti_j is the working efficiency of the agent represented by $AToken_j$, Loc_j is the current location of the agent represented by $AToken_j$ and MSp_j represents the moving speed of the agent represented by $AToken_j$.

An agent A_j can have two different roles, a *coordinator* or/and a *resource provider* (an agent can be a coordinator, a resource provider or both).

Definition 5. A *Coordinator* is an agent, which is in charge of allocating agents to tasks.

Definition 6. A *Resource Provider* is an agent, which has the resource requested by a task and can finish the task before the deadline.

Since communication range limitation is a common problem in disaster domains, we assume that agents can only discover the information about tasks nearby and communicate with other agents close to its location within a limited communication range.

Definition 7. A *Neighbour* of A_j is an agent that can directly communicate with A_j. In other words, a neighbour of A_j is an agent within A_j's communication range, (i.e. $Comm_j$ see Definition 2).

3   The Basic Processes of Our Approach

Our approach consists of five looped steps, which are (1) task token creation, (2) group formation, (3) token passing, (4) task allocation and (5) allocation solution return. After one loop of five steps, if there are available agents and tasks in the domain, these five steps will be repeated again according to the current location of available tasks and agents.

3.1   Task Token Creation

In this step, each available agent A_j (i.e. $ASta_j$ = "available", see Definition 2) checks the area within its communication range ($Comm_j$). Once, A_j finds an available task T_i (i.e. $TSta_i$ = 'available', see Definition 1), A_j creates a task token, $TToken_i$ =< A_j, T_i, $DLine_i$, $WLoad_i$, Loc_i, Emg_i > (see Definition 3) for T_i. Meanwhile, A_j also creates an agent token $AToken_j$ (see Definition 4) for itself.

3.2   Group Formation

After task tokens are created, agents will form groups and are allocated tasks within groups. Due to communication range constraints, agents can only communicate with their neighbours (see Definition 7). In order to make decisions for task allocation based on available resources as much as possible, the group formation mechanism is designed to connect as many agents as possible within the domain. The objective of the group formation mechanism is to link agents with their neighbours and form a tree, in which the agent at the root node is chosen to be the coordination of the group. In the mechanism, we defined three "neighbour related" variables for each agent, including, the parent agent (PA), the coordinator (C) and the number of neighbors of the coordinator (NNC). For example, for an agent A_j, the three variables of A_j is denoted as A_j.PA, A_j.C and A_j.NNC, respectively. The group formation mechanism is conducted based on an iteration process.

The group formation mechanism is described by Algorithm 1. Before iteration, the three variables of each agent (i.e. A_j) are initialised as follow: A_j.PA is set to A_j, A_j.C is set to A_j and A_j.NNC is set to the number of neighbours of A_j (since A_j is the coordinator of itself) (see line 2). Each agent (e.g. A_j) repeats the following three steps. **Step 1:** A_j finds an agent (e.g. A_u) which has the highest value of the variable NNC (i.e. A_u.NNC) among neighbours of A_j (including A_j itself) (see line 5); **Step 2:** the variables of A_j (A_j.PA, A_j.C and A_j.NNC) are updated as follow: A_j.PA is set to A_u, A_l.C is set to A_u.C and A_j.NNC is set to A_u.NNC (see line 6); **Step 3:** After above two steps, each neighbour of A_j (e.g. A_l), compares its coordinator (A_l.C) with the coordinator of A_j (A_j.C). If a different coordinator

---

Algorithm 1 Group Formation

1: **for each** Agent A_j **do**
2:    A_j.PA$\leftarrow A_j$; A_j.C$\leftarrow A_j$; A_j.NNC$\leftarrow$ number of neighbours of A_j
3:    **for each** Agent A_j **do**
4:      Get A_u from A_j and its neighbours, where A_u.NNC is Maximum A_j.PA$\leftarrow A_u$; A_j.C$\leftarrow A_u$.C; A_j.NNC$\leftarrow A_u$.NNC
5:       **for each** neighbours of A_j:A_l **do**
6:        **if** A_l.C$\neq A_j$.C **then**
7:         A_l.PA$\leftarrow A_j$; A_l.C$\leftarrow A_j$.C; A_l.NNC$\leftarrow A_j$.NNC
8:        **end if**
9:       **end for**
10:    **end for**
11: **end for**

---

has been found (A_l.C $\neq A_j$.C), the variables of A_l (A_l.PA, A_l.C and A_l.NNC) are updated as follow: A_l.PA is set to A_j, A_l.C is set to A_j.C and A_l.NNC is set to A_j.NNC (see lines 8–9). The above three steps (lines 4–9) will be repeated by each agent until no further updating for three variables of any agent.

3.3 Token Passing

In this step, each agent A_j first gets task tokens (see Definition 3) and agent tokens (see Definition 4) (created in the task token creation step) from its child agents. Different agents in the same group might create task tokens for the same task. For example, A_u and A_l are child agents of A_j and they create task tokens $TToken_i =< A_u, T_i, DLine_i, WLoad_i, Loc_i, Emg_i >$ and $TToken_i =< A_l, T_i, DLine_i, WLoad_i, Loc_i, Emg_i >$ for the same task T_i. It is can be seen that except for the first element (i.e. the agent who created the token) the two task tokens created by A_u and A_l are exactly the same. In this situation, A_j combines the two tokens together ($TToken_i$ created by A_l is abandoned and only $TToken_i$ created by A_u is kept). After this, A_j passes the combined task token/tokens and agent tokens that A_j received and $AToken_j$ to its parent agent. Finally, all of task tokens and agent tokens are passed to the coordinator of the group.

Figure 1 shows an example of token passing. There are four agents: A_1, A_2, A_3 and A_4, in the example. A_1 is the parent agent of A_2 (A_2.PA=A_1), A_2 is the parent agent of A_3 and A_4 (A_3.PA=A_2 and A_4.PA=A_2). There are three tasks in the example: T_1, T_2 and T_3, T_1 and T_2 can be found by A_3 and T_2 and T_3 can be found by A_4. After the task token creation step, A_3 creates two task tokens (see, Definition 3) for T_1 and T_2, which can be represented as "$< A_3, T_1, \ldots >$" and "$< A_3, T_2, \ldots >$", respectively, while A_4 also creates two task tokens (see Definition 3) for T_2 and T_3, which can be represented as "$< A_4, T_2, \ldots >$" and "$< A_4, T_3, \ldots >$", respectively. Then, the task tokens and the agent tokens (see Definition 4) of A_3 and A_4 are passed to A_2. After checking the task tokens getting from A_3 and A_4, A_2 discovers that task

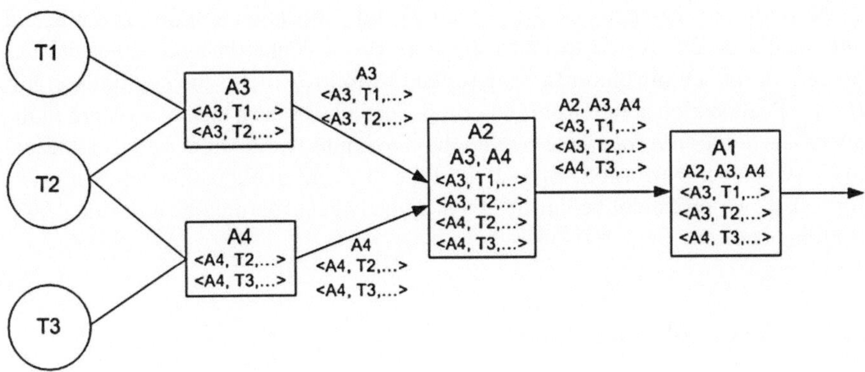

Fig. 1 The example of token passing

tokens "$< A_3, T_2, \ldots >$" and "$< A_4, T_2, \ldots >$" are created for the same task by two different agents. Therefore, A_2 combines the task tokens "$< A_3, T_2, \ldots >$" and "$< A_4, T_2, \ldots >$" to "$< A_3, T_2, \ldots >$". After combination of task tokens, A_2 passes its agent tokens (A_2, A_3 and A_4) and the combined task token ("$< A_3, T_2, \ldots >$") and its other task tokens ("$< A_3, T_1, \ldots >$" and "$< A_4, T_3, \ldots >$") to A_1.

3.4 Task Allocation

This step is the core of the approach. In this step, a coordinator tries to find the best allocation solution of its group according to available tasks and agents in the group. In our approach, the allocation solution of a group is represented by an n-tuple $Alloc =< ANo_1 \rightarrow TNo_1, ANo_2 \rightarrow TNo_2, \ldots, ANo_n \rightarrow TNo_n >$, where, ANo_j is the ID of the jth agent of the group and TNo_j is the ID of the task to which the jth agent of the group is allocated. If the jth agent of the group is not allocated to any task, the TNo_j is $\varnothing$. Since finding the best allocation solution is an NP-hard problem (the proof process can be found in [11]), our approach employs heuristics. The heuristics for finding the best allocation solution can be implemented in two sub-steps (1) the elimination of unuseful allocation solutions and (2) utility calculation.

3.4.1 The Elimination of Unuseful Allocation Solutions

Based on the combinatorics, the number of combination for a group with m number of tasks and n number of agents is $(m + 1)^n$. Among these allocations, there are allocations where the allocated agents cannot finish the task in time (before deadline). For example, in a group, one of allocation solution is $Alloc_a =< A_1 \rightarrow T_1, A_2 \rightarrow T_2, A_3 \rightarrow T_1 >$. After calculation, if we find that A_1 and A_3 cannot finish

T_1 according to the status of A_1, A_3 and T_1, this allocation solution is not useful and should be eliminated to make the following calculation more efficient. The process of this elimination can be described as follows. First, the coordinator gets $(m + 1)^n$ allocation solutions of the group. Second, for each task T_i existing in the allocation solution *Alloc*, we calculate the maximum workload of agents allocated to T_i, which can perform before the deadline of T_i. If an allocation solution exists for task T_i which cannot be finished by allocated agents before deadline, the $Alloc_a$ is eliminated.

3.4.2 The Calculation of Utility

In this sub-step, the coordinator calculates the utility of each useful allocation solution and chooses the allocation solution with the highest utility value as the best allocation solution for the group. To judge an allocation solution depends ont only on how many tasks that can be finished, but also on the benefits received from finishing the tasks and the cost spent on finishing the tasks. In our approach, the utility of an allocation solution $Alloc_a$ is affected by two factors, which are (1) the benefits received from finishing the task and (2) the cost of finishing the tasks.

With the two factors above, the utility of an allocation solution $Alloc_a$ can be calculated by the following formula.

$$Utility_{Alloc_a} = Q_{benefit} - Q_{cost}. \tag{2}$$

The benefits received from finishing tasks involves two factors: (1) the weighted workload of each finished task and (2) the time saved by finishing each task. The benefits of an allocation solution *Alloc* is the sum of the benefits of all finished tasks in $Alloc_a$, which is calculated as follows.

$$Q_{benefit} = \sum_{t=0}^{n} Emg_i \cdot \left(WLoad_i + \frac{(DLine_i - FTime_i) \times WLoad_i}{DLine_i} \right), \tag{3}$$

where Emg_i, $WLoad_i$, $DLine_i$ are the degree of urgency, the workload and the deadline of T_i in $Alloc_a$ and $FTime_i$ is the predicted finishing time of T_i in $Alloc_a$.

The cost spent on finishing tasks in our approach includes (1) the traveling time of each agent moving from its current location to the location of the its allocated task and (2) the time each agent spends on finishing its allocated task. Therefore, the cost of an allocation solution $Alloc_a$ is the sum of the cost of agents in $Alloc_a$, the cost of traveling and of finishing tasks, which can be calculated by Eq. (4) as follows.

$$Q_{cost} = \sum_{t=0}^{m} ((FTime_i - CurTime) \times Uti_j), \tag{4}$$

where Uti_j is the working efficiency of A_j, $FTime_i$ is the predicted finishing time of T_i in $Alloc_a$, to which A_j is allocated and $CurTime$ is the current time when the coordinator calculates the utility function of $Alloc_a$. $FTime_i$ can be calculated as follows:

$$FTime_i = \frac{WLoad_i + CurTime \times \sum_{j=1}^{n} Uti_j + \sum_{j=1}^{n} \frac{Dis(T_i,A_j)}{MSp_j} \times Uti_j}{\sum_{j=1}^{n} Uti_j}, \tag{5}$$

where $Wload_i$ is the workload of T_i in $Alloc_a$, $CurTime$ is the current time point when the coordinator calculates the utility function of $Alloc_a$, Uti_j is the working efficiency of A_j, which is allocated to T_i in $Alloc_a$, $Dis(T_i, A_j)$ is the distance between the current location of A_j to the location of T_i and MSp_j is the move speed of A_j.

3.5 *Allocation Solution Return*

In this step, the coordinator sends the best allocation solution to the agents in the group. Once an agent received the allocation solution, it will do the following things. First, the tokens of allocated tasks (which it can definitely finish) is eliminated from the agent token pool. Second, the allocation is passed to the neighbours of the agent and the status of the agent is changed from "available" to "working". Finally, the agent moves to the location of the task and starts working on it. When the first agent reaches the location of a task, it changes the status of the task from "available" to "working". After finished the task, all of agents, which worked on the task, change their status from "working" to "available" and the status of the task is also changed from "working" to "finished".

4 Experiment and Analysis

Two experiments are conducted to evaluate the performance of our approach. Experiment 1 is to evaluate the performance of our approach under different communication range of agents. The benchmark of Experiment 1 is the F-max-sum algorithm proposed by Ramdchurn et al. [10]. The F-max-sum is a decentralised task allocation mechanism. Experiment 2 is to evaluate the impact of degree of urgency of task on the performance of our approach. Two experiments are demonstrated and analysed in detail in the following two subsections, respectively.

4.1　Experiment 1

The purpose of this experiment is to test the impact of the different communication ranges on the performance of the proposed approach.

4.1.1　Experiment Setting

In Experiment 1, 100 tasks (in which 50 tasks already exist at the beginning of the experiment and 50 new tasks are discovered between 0 and 100 time units) and 10 agents with different initial locations in a 50×50 area are employed. Since the degree of urgency of each task is not being tested in Experiment 1, the degrees of urgencies of all tasks are set to 1. In addition, all tasks have a workload between 10 and 60 work units and must be finished before the deadline between 5 and 100 time units. Agents can only finish 1 work unit or move 1 distance per time unit. The settings of Experiment 1 are shown in Table 1.

In this experiment, we use three scenarios to represent three different kinds of communication ranges of agents, shown in Table 2.

This setting tries to simulate tree different communication conditions, i.e. "isolated", "limited" and "full" communication. The two approaches (i.e. the proposed approach and the benchmark approach) use their utility functions to choose which tasks should be finished and which tasks should be abandoned. In addition, the Manhattan distance [14] is employed to calculate the distance between two locations in the experiment. The group formation mechanism is affected by three factors: the number of agents in the domain, the communication range of agents and the size of the domain. From the definition of the Manhattan distance, in an $m \times m$ domain, if the communication range of agent A_j is c, the possibility that another agent is the neighbour of A_j is calculated as follows.

$$Possibility = \frac{2 \cdot c^2 + 2 \cdot c + 1}{m \cdot m} \tag{6}$$

Table 1　The settings in Experiment 1

Name	Value	Name	Value	Name	Value
Number of tasks	50	Deadline of tasks	$5 \sim 100$	Number of agents	10
Number of new tasks	50	Workload of tasks	$10 \sim 60$	Work efficiency	1
Urgent degree	1	Area size	50×50		

Table 2　Three scenarios in Experiment 1

Scenario	Communication range (unit)	Communication capacity
1	10	Isolated
2	20	Limited
3	50	Full

With increase of communication range or agent numbers and decrease of domain size, more agents can find neighbours and form groups.

Since the F-max-sum does not consider the communication range, we assume that the communication ranges of agents in both approaches are limited. In most situations, agents need to cooperate to complete tasks. The two approaches are compared with three different communication ranges between agents 10, 20 and 50 in the three scenarios, respectively. In Scenario 1 (10 communication range), according to Eq. (6), in an 50×50 area, the possibility of two agents being neighbour of each other is 8.84 %, which is an extreme "lonely" setting for 10 agents. In Scenario 2 (20 communication range), the possibility of two agents being neighbour of each other is 33.6 %. This is a relatively good setting for 10 agents, because there is increased possibility for agents to form groups and there will also be some agents that cannot connect with other agents. In Scenario 3 (50 communication range), there is no communication range constraints between agents and all agents can communicate with each other directly.

4.1.2 Experimental Results and Analysis

The experimental results for three scenarios are shown in Figs. 2, 3 and 4, respectively. The X-axis of three figures represents the consumed time units. The Y-axis of three figures indicates the number of tasks that have been finished.

Scenario 1: Since the communication range for each agent is only 10, in this scenario, it is hard for agents to work cooperatively (i.e. agents are isolated). All of the agents in both approaches work independently. From Fig. 2, we can see that there are nearly the same finished workloads between the two approaches.

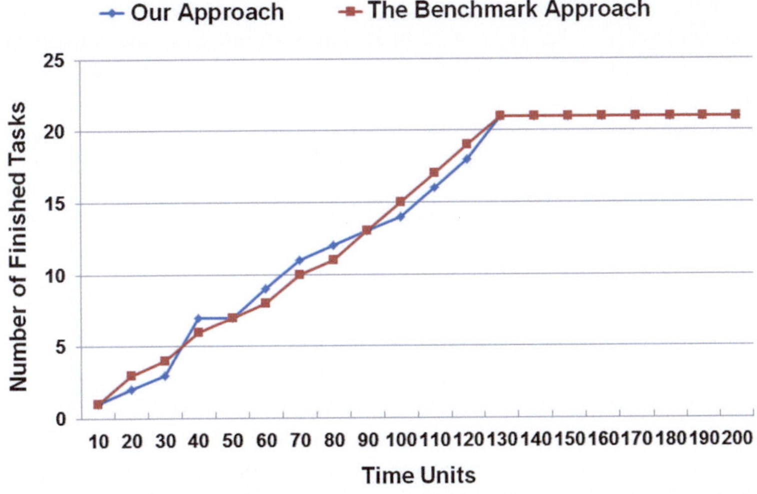

Fig. 2 Experimental result in Scenario 1

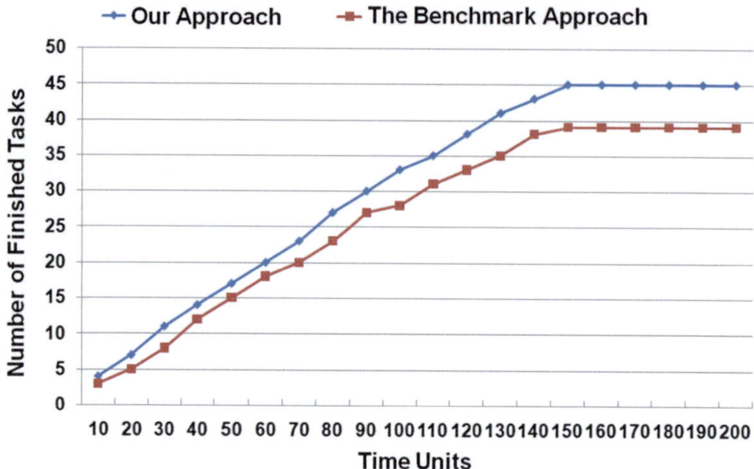

Fig. 3 Experimental result in Scenario 2

Scenario 2: The communication ranges of agents are 20, in this scenario, and according to Eq. (6), the probability for two agents in the domain to be neighbours is around 33 %, so the communication capacity is limited. From Fig. 3, we can see that our approach has better performance than the benchmark approach. That is because when agents form groups, the coordinator in our approach can collect the task and agent information from the members of the group and find the most suitable allocation solution for the group. However, the benchmark approach is limited by communication range and agents cannot have full knowledge for all of tasks. Hence, agents can not offer the comprehensive messages to other agents and this affects the decision making of agents to achieve good performance in the domain.

Scenario 3: The communication range of this scenario is 50 in a 50×50 area, so an agent can have a global view of all tasks and other agents in the domain. Agents in the domain can fully communicate with any other agents. Figure 4 demonstrates that our approach is equal to a centralised approach and the F-max-sum algorithm can also achieve the highest efficiency with the full communication capacity, so the performance of the two approaches is nearly the same.

From the experimental results produced from three scenarios, it can be seen that the communication range has a bigger impact on the F-max-sum algorithm for task allocation than that of on our approach. The reason is that the F-max-sum algorithm requests good communication capacity between agents, so decision making for each agent in an area is limited if an agent can not offer comprehensive information of itself to other agents. In our approach, since the group formation mechanism can help the coordinator to find much information in a domain, the coordinator can reduce the impact of the limited communication range on task allocation and make the most suitable decision for the group. Therefore, our approach performs better than the benchmark approach when there is limited communication range.

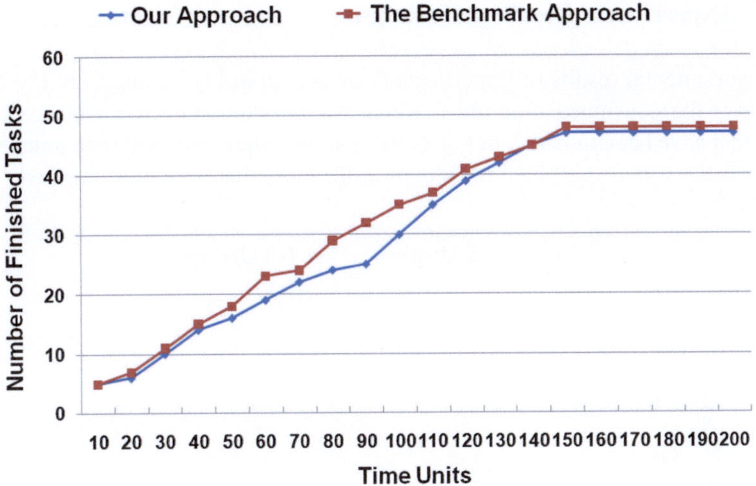

Fig. 4 Experimental result in Scenario 3

4.2 Experiment 2

The purpose of this experiment is to test the impact of different degrees of urgencies of tasks on task allocation.

4.2.1 Experiment Settings

In Experiment 2, 100 tasks and 10 agents with different initial locations in a 50×50 domain are employed. All of the tasks have workloads between 10 and 60 work units, which must be finished before the deadline between 5 and 100 time units. Agents in the domain can only finish 1 work unit or move 1 distance per time unit and have a fixed communication range of 20. To evaluate the impact of degree of urgency, our approach is conducted twice. First time, the degrees of urgencies of all tasks are 5, which indicate that there is no difference between the degrees of urgencies of tasks. In the second time, the degrees of urgencies of all tasks varied between 1 and 9 with normal distribution. The settings of the experiment are shown in Table 3.

Table 3 The settings in Experiment 2

Name	Value	Name	Value	Name	Value
Number of tasks	100	Workload of tasks	$10 \sim 60$	Work efficiency	1
Urgent degree	$1 \sim 9$	Area size	50×50	Communication range	20
Deadline of tasks	$5 \sim 100$	Number of agents	10		

4.2.2 Experimental Results and Analysis

The experimental results of Experiment 2 are shown in Figs. 5 and 6. In Fig. 5, the X-axis is the consumed time units, while Y-axis is the sum of workloads of the tasks that have been finished. In Fig. 6, the X-axis is the consumed time units, while Y-axis is the sum of weighted workloads of the tasks that have been finished.

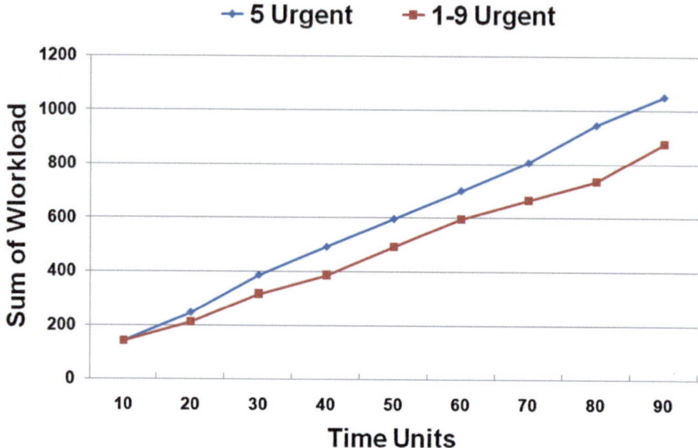

Fig. 5 Sum workload

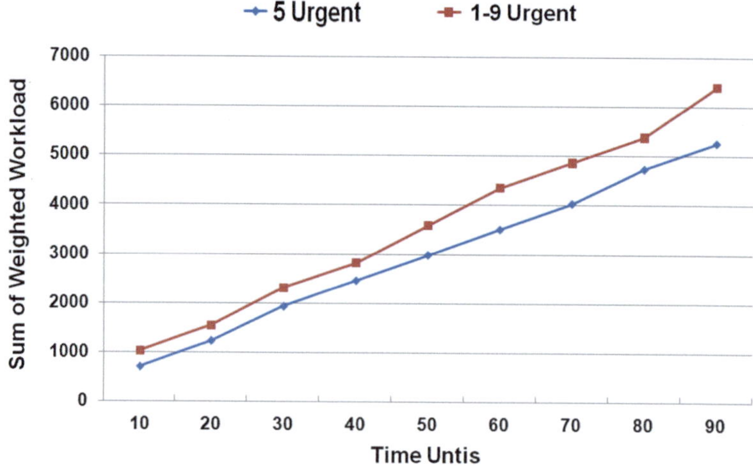

Fig. 6 Sum weighted workload

From Figs. 5 and 6, it can be seen that the proposed approach can finish more workload, when the degrees of urgencies of tasks are equal. This is because when the degrees of urgencies are similar for all tasks, the coordinator allocates agents to tasks only according to the workload of each task and the distance between agents and tasks. However, the proposed approach can finish more weighted workload, when the degrees of urgencies of the tasks are different. The reason for this situation is that the proposed approach consider the degrees of urgencies of tasks in its utility calculation function. Therefore, if we differentiate the degrees of urgencies of tasks in the domain, the proposed approach can allocate the more urgent tasks before the less urgent tasks, which is reasonable.

4.3 Summary

From Experiments 1 and 2, we can conclude that (1) the communication range of agents and the degrees of urgencies of tasks have great impact on decision making for task allocation, particularly in a diaster domain, because the limited communication range reduces the local view of an agent; and (2) using the equivalent the degrees of urgencies for all tasks can bury the importance of tasks.

Our approach uses group formation mechanism to expand the local view of agents and differentiates tasks by employing the concept of the degree of urgency of each task during task allocation. The experimental results indicate that our approach has better performance than the famous approach (F-max-sum) under limited communication range. In addition, using the weighted workload concept can help agents to finish more urgent tasks under time constraints.

5 Related Work

Ramamritham et al. proposed a distributed coordination approach to handle task allocation with deadline and resource requirements in [9]. Their approach solves common task allocation problem based on the classification of tasks. The weakness of their approach is the it omits the space and communication constraints in the process of task allocation. Furthermore, in a disaster domain, most of tasks are critical and few of them are other categories. Therefore, the approach proposed by Ramamritham et al. is hard to be applied to a disaster domain. Our approach considers the space and communication constraints and can also deal with tasks with different degrees of urgencies during task allocation in a diaster domain.

Some researchers such as Katirya and Jennings have used Mixed Integer Linear Programming (MILP) [4] to get the optimal allocation solution in disaster domains in [5, 11]. The MILP process can guarantee an optimal solution for task allocation if the coordinator have a global view of the domain. However, the MILP-based approach has three main weaknesses. First, in a common disaster domain, it is hard for a coordinator to have a global view due to space and communication constraints. Second, the MILP can only deal with problems with fixed numbers of tasks and agents. These features make their solutions hard to fit the highly dynamic natures of disaster domains. Third, the computation complexity of the MILP is high. Our approach can solve dynamic task allocation problems in a decentralised manner.

In recent years, the DARPA coordinators program is a popular environment for the coordination of task allocation, which has a complicated hierarchy task structure. The difference between the DARPA coordinators program and ours can be summarized as follows. (1) The tasks in the DARPA coordinators program do not have clear deadlines. (2) Most proposed models [1,7,8,13] dealing with the DARPA coordinators program do not consider either space or communication constraints. The approach proposed in this paper can handle the coordination problem in disaster domains with both space, time and communication constraints.

> **Conclusion and Future Work**
>
> In this paper, a novel decentralised coordination approach for task allocation in disaster domains with space, time and communication constraints is proposed. The approach uses a comprehensive utility function to enable a coordinator to find the best solution for task allocation in the group by considering degree of urgency of tasks, workload and traveling time of agents to the location of a task under multiple constraints. A group formation mechanism is set up in order to help agents to form groups under limited communication range due to the destruction of infrastructures in a diaster domain. In addition, the degree of urgency of each task and dynamic features of a such domain are covered by our approach in order to meet the requirements of disaster domains. The experimental results have shown that our approach can provide a more efficient way to meet the challenges of task allocation in a disaster domain than other existing approaches. In the future, we will work on task allocation problems in more complex environments (e.g., agents can only have spars interaction).

References

1. Barbulescu, L., Rubinstein, Z.B., Smith, S.F., Zimmerman, T.L.: Distributed coordination of mobile agent teams: the advantage of planning ahead. In: AAMAS '10, pp. 1331–1338, Richland (2010)
2. Chapman, A.C., Micillo, R.A., Kota, R., Jennings, N.R.: Decentralised dynamic task allocation: a practical game: theoretic approach. In: AAMAS '09, vol. 2 (2009)
3. Farinelli, A., Rogers, A., Petcu, A., Jennings, N.R.: Decentralised coordination of low-power embedded devices using the max-sum algorithm. In: AAMAS '08, vol. 2, pp. 639–646, Estoril (2008)

4. Gordon, G.J., Hong, S.A., Dudík, M.: First-order mixed integer linear programming. In: UAI '09, pp. 213–222. AUAI Press, Arlington (2009)
5. Koes, M., Nourbakhsh, I., Sycara, K.: Heterogeneous multirobot coordination with spatial and temporal constraints. In: AAAI '05, pp. 1292–1297. AAAI Press, Menlo Park (2005)
6. Lesser, V.: Cooperative multiagent systems: A personal view of the state of the art. IEEE Trans. Knowl. Data Eng. **11**, 133–142 (1999)
7. Maheswaran, R., Szekely, P., Becker, M., Fitzpatrick, S., Gati, G., Jin, J., Neches, R., Noori, N., Rogers, C., Sanchez, R., Smyth, K., Buskirk, C.V.: Look where you can see: Predictability & criticality metrics for coordination in complex environments. In: AAMAS '08 (2008)
8. Musliner, D.J., Goldman, R.P.: Coordinated plan management using multiagent mdps. In: AAAI '06, pp. 73–80. AAAI Press, Menlo Park (2006)
9. Ramamritham, K., Stankovic, J.A., Zhao, W.: Distributed scheduling of tasks with deadlines and resource requirements. IEEE Trans. Comput. **38**(8), 1110–1123 (1989)
10. Ramchurn, S.D., Farinelli, A., Macarthur, K.S., Jennings, N.R.: Decentralized coordination in robocup rescue. Comput. J. **53**(9), 1447–1461 (2010)
11. Ramchurn, S.D., Polukarov, M., Farinelli, A., Truong, C., Jennings, N.R.: Coalition formation with spatial and temporal constraints. In: AAMAS 2010, pp. 1181–1188 (2010)
12. Ren, F., Sim, K.M.: Adaptive conceding strategies for automated trading agents in dynamic, open markets. Decis. Support Syst. **48**(2), 331–341 (2009)
13. Smith, S.F., Gallagher, A., Zimmerman, T.: Distributed management of flexible times schedules. In: AAMAS '07, vol. 74, pp. 1–8. ACM, New York (2007)
14. Taxicab geometry (2003). https://en.wikipedia.org/wiki/Centroid
15. Vinyals, M., Pujol, M.: Divide-and-coordinate: Dcops by agreement. In: AAMAS'10, pp. 149–156 (2010)

Describing and Evaluating Assistance Using APDL

Martin Nyolt, Alexander Steiniger, Sebastian Bader, and Thomas Kirste

Abstract We introduce an approach to test and evaluate assistance systems for smart environments. Our work comprises a specification language to describe assistance problems independent of the concrete application domain and a test system to test assistance systems and compare their performance systematically. For the latter, our test system employs software-in-the-loop simulation to test assistance systems in a virtual environment instead of a real deployment. On the one hand, this helps developers of assistance systems to evaluate their implementations and compare them to other systems. On the other hand, the language allows researchers specifying assistance problems at an abstract level. Furthermore, we provide a formal semantics for our language, describe the usage of our test system, and show its applicability using a simple assistance system.

Keywords Assistance systems • Formal language • Simulation • Smart environments

1 Introduction and Motivation

The development of *smart environments* (or *intelligent environments*) has been an active research area for the last two decades. The area covers various sub-disciplines of computer science ranging from artificial intelligence (AI) over middleware systems to human computer interaction. Good introductions and overviews are available [5, 25, 29]. Prominent examples of larger installations are the *MavHome* [6], the *Gator Tech Smart House* [13], the *Georgia Tech Aware Home* [19], and the *Adaptive House* [22]. But also other application domains have been addressed, e.g., intelligent classrooms [8], team meeting rooms [1], command and control rooms [3], and domestic environments [4].

One fundamental aspect of smart environments is to assist their users proactively [5], often by integrating AI techniques. This is done by an—what we

M. Nyolt (✉) • A. Steiniger • S. Bader • T. Kirste
University of Rostock, Rostock, Germany
e-mail: martin.nyolt@uni-rostock.de; alexander.steiniger@uni-rostock.de; sebastian.bader@uni-rostock.de; thomas.kirste@uni-rostock.de

© Springer Japan 2015

Q. Bai et al. (eds.), *Smart Modeling and Simulation for Complex Systems*, Studies in Computational Intelligence 564, DOI 10.1007/978-4-431-55209-3_5

call—*assistance system* (AS), which analyses the users' activities and interactions, infers the users' intentions and goals, and derives "useful" actions that can be executed automatically to support the users while they are trying to achieve their goals in a smart environment. We call the problem of determining those actions in a certain scenario an *assistance problem*. In this regard, describing assistance problems formally, if possible independent of the actual application domain while taking the characteristics of smart environments into account, is a challenge.

Since applications running in smart environments, such as ASs, influence the convenience or even safety of their users, those applications need to be tested [26]. Testing high-level ASs in general and evaluating and comparing their performance are other challenges. Conducting tests in real environments is difficult and expensive [26] and it can be extremely difficult to reproduce testing conditions [18], especially when humans are involved. Thus, we use simulation to test ASs in a systematic and controlled manner. For this, an AS under test is "plugged into" a virtual environment, which emulates the physical target environment. Thereby, we merely support black-box testing [21] of ASs. Other tests, such as white-box tests or static program analysis, are not in the scope of our work.

We introduce a test system that comprises a formal language for describing assistance problems—the *Assistance Problem Definition Language* (APDL)—and that interoperates with a suitable simulation system, which is used to simulate smart environments. Our test system can be used to test ASs individually and for comparing the performance of different ASs regarding specific assistance problems. We believe that both testing and evaluating can profit from a formally specified language and that in particular that latter is crucial for a systematic development of ASs. As such, we present the concepts of how we use simulation for testing assistance systems.

2 Related Work

The benefits of testing and evaluating assistance systems based on simulation have already been discussed in the literature. Huebscher and McCann [18] simulate sensor data, especially activity and location information, based on which context recognition tools can be tested. Krger et al. [20] generate sensor data of varying quality (correctness of the data) by means of simulation. Based on that data, the robustness of a probabilistic activity recognition approach is systematically tested and evaluated. Similarly, *Persim* simulates human activities and sensors of smart environments to evaluate activity recognition algorithms [14]. Proetzsch et al. [30] developed a language to describe test cases for autonomous mobile robots and introduce a system to execute those test cases automatically. For testing real off-road robots, *in-the-loop simulation* is employed. Gierke et al. [11] present a model-based approach to test the functionality and timeliness of the assistance system *Autominder* by coupling it with a virtual environment, which includes a user model.

Listing 1 Simple PDDL action definition, moving an object o from position x to y.

```
(:action move :parameters (?o ?x ?y)
  :precondition (at ?o ?x)
  :effect (and
            (not (at ?o ?x))
            (at ?o ?y))
)
```

In addition to approaches that focus on testing ASs individually, competitions exist in which the performance of different systems or algorithms can be compared regarding different test scenarios. The EvAAL competition[1] focuses on specific sub-problems (localisation, activity recognition) ASs have to solve. Another central problem in smart environments is planning, i.e. select an (optimal) sequence of actions to support users of smart environments, also known as strategy generation [12]. In the International Planning Competition[2] (IPC), planning algorithms are evaluated for different problem types and domains. Domain knowledge is not encoded in the planning algorithms, but a machine-readable description of the problem is given as input during the contests. Accordingly, any appropriate planning tool can be used for solving arbitrary problems defined in such domain-independent languages. The IPC established the Planning Domain Definition Language (PDDL) [10] as a well-known and widely used language for defining all kinds of planning domains and problems.

PDDL defines a set of actions, each with preconditions and effects over first order predicates on the world state. Listing 1 gives an example of such a typical PDDL action. PDDL problem instances define a set of initial predicates and goal predicates and ask for a sequence of actions reaching a goal state. One central assumption in PDDL is that actions are only executed by the planning agent. Autonomous behaviour of the modelled world is not expressible. That is, the agent is of full and exclusive control over the world. There are no other causes of or influences on the world changing its state. Incomplete observations and non-deterministic action outcomes are supported in later versions of PDDL. The Relational Dynamic Influence Diagram Language (RDDL) [31], another problem definition language that is partly based on PDDL, supports autonomous world dynamics from the outset. In contrast to PDDL, state transitions are independently executed as part of the model semantics itself. Actions executed by the agent are only one input variable state transitions may depend upon. RDDL has also been designed to specify planning problems with uncertainty (noisy observations and actions).

A different approach for modelling systems with uncertainty is using speci-fications based on statistical relational learning. One example is *PRISM* [32], a logic-based probabilistic modelling language. Usually, domain models have to be specified by hand. As a tool, PRISM provides the ability to partially learn model

---

[1]EvAAL competition homepage: http://evaal.aaloa.org

[2]IPC homepage: http://ipc.icaps-conference.org

parameters (e.g., sensor behaviour) automatically from data without human experts. Another example is the rational language *iBal* [28]. It does not only allow defining action semantics, it also supports decision making. A reward is specified for actions and the inference engine can select the most valuable action. This way, iBal may also be used as an AS itself.

3 Requirements Analysis

We extend previous work by using simulation of smart environments to test ASs against complete assistance problems, not only sub-problems thereof. For this, we do not implement a new simulation system (or formalism), but aim to use an existing one. Analogous to PDDL and other languages, we also need a general-purpose assistance problem description language. This language shall be used by AS to reason about its current assistance problem.

In the following, we identify requirements on (a) simulation for being used to test and evaluate ASs and (b) languages to describe assistance problems formally. The previous covers requirements on a suitable formalism to define executable simulation models[3] and on the execution of these models, on a corresponding simulation system.

3.1 *Simulation Requirements*

As we want to test ASs in simulated smart environments instead of real ones, *the formalism used to define executable simulation models of smart environments and the simulation system on which these models are executed have to allow the communication and interaction between an external process—the AS under test—and the simulation.* Thereby, actions of the AS affect the simulated smart environment similarly as they would affect the real smart environments and vice versa. Furthermore, the time (wall-clock time) consumed by the AS has to be related to and converted into simulation time. In short, we want to employ software-in-the-loop simulation [2]. However, it should be transparent to the AS under test whether it is embedded in a real or a simulated smart environment.

ASs operate on data gathered by the sensors deployed in smart environments rather than the actual world state, i.e. the ground truth. However, for an evaluation of ASs this ground truth, e.g., the exact positions of users or their current activities, is needed. Thus, *all information that is required for the evaluation—the ground truth—as well as the information provided by sensors have to be captured in the simulation model, i.e. its state.* But to ensure a realistic setting, *only an explicitly*

---

[3]In the simulation, a smart environment is represented by a model: the simulation model.

defined part of this information should be accessible to the AS under test. For computing actual evaluation data, *the simulation system must allow extracting the state of the simulation model—which serves as ground truth—during the simulation.* Note that we distinguish between the extracted ground truth and the data that is communicated between an AS and the simulation.

The sensor data ASs have to cope with are often noisy and intermittent. Moreover, devices and users may exhibit a non-deterministic behaviour. Therefore, *it must be possible to employ discrete and continuous probabilistic effects in the simulation.*

3.2 Language Requirements

A formal description of an assistance problem is needed for two purposes: First, the test and evaluation system needs a model to simulate. A specification of test cases is also mandatory for comparable and reproducible results. Second, the AS itself needs some information about the environment it is deployed in.

Assistance systems may be tailored to specific smart environments and use cases, or they may be more application-independent. For instance, an AS may only support lectures in rooms with a special configuration and equipment. A more elaborate AS may support other environments with a different configuration and more use cases. As a result, some ASs may need more or less information about the environment they are deployed in than others. To create suitable descriptions for different kinds of ASs within our test system, *the language must be extensible for providing the necessary level of detail for the AS.*

In order to evaluate and compare ASs, *a reward function must be part of each problem specification.* The reward function is used as a performance measure for the simulation and is also intended to be used by more advanced ASs for deciding which actions to execute. In our work, we define *performance* to be a level of quality of support that can be assigned to an assistance system. The reward is in no way a thorough value of performance, but only one possible indicator. As the reward function is domain-dependent, it must be defined specifically for the domain *and* intended testing purpose. For instance in a smart meeting room presentation, perceived assistance performance may differ between the audience and the lecturer.

Often, devices and users of an smart environment act independently of each other. To make use of this independence in reasoning and support modular design, *the language should facilitate the definition of objects, their properties, and their behaviour.* The behaviour of an object is usually influenced by internal events (autonomous behaviour) or external stimuli from other objects. *The language must allow to describe both autonomous and externally triggered behaviour.*

ASs must reason about received events and possible actions. But neither all events are public in the sense that they should be accessible by the AS, nor all actions are executable. In particular, internal state changes of some object or actions use to formalise human behaviour should not be accessible for the AS. Thus, *it must be possible to declare observable/hidden properties and actions that can/cannot be executed by the assistance system.*

3.3 Discussion of Existing Systems and Languages

To the best of our knowledge, no approach exists to measure and evaluate the overall performance of ASs from analysing sensor data to decision making. Existing competitions such as EvAAL, IPC, or the General Game Playing [9] served as inspiration while developing our framework, but none satisfied our needs completely. EvAAL concentrates on certain aspects of assistance only, in particular planning, sensing, and activity recognition. The IPC focuses on planning only and the General Game Playing contest does neither allow specifying probabilities nor of fine-grained reward functions.

The inspected simulation-based approaches [14, 18] focus on testing ASs individually and do not allow an interaction between the simulation and the actual system under test, instead synthetic sensor data is generated and used for testing afterwards. The modular, hierarchical simulation formalism PepiDEVS (Parallel External Process Interface Discrete Event System Specification) [15] for discrete event simulation allows defining complex simulation models that can communicate with external processes, e.g., ASs, in a well-defined manner. PepiDEVS exists as a plugin for the plugin-based, general-purpose simulation and experimentation framework JAMES II [16]. This and the availability and flexibility of JAMES II turn the framework into a suitable candidate for a simulation system with which our test system can interoperate. Together PepiDEVS and JAMES II satisfy all our requirements to the simulation (see Sect. 3.1). The applicability of combining JAMES II and the ancestor of PepiDEVS has already been demonstrated [11, 33]. However, also other suitable simulation systems and formalisms may be used within our test system.

Unfortunately, none of the languages mentioned in Sect. 2 fulfils all our requirements to such a language. PDDL does not allow modelling autonomous behaviour. RDDL supports probabilistic effects and a reward function. However, RDDL intrinsically does not support continuous time and durative actions. In addition, both languages do not support a modular model design.

4 Presentation of the Test System

The test system we propose here addresses two objectives: (a) test and evaluate the implementation of a single AS for specific assistance problems and (b) compare the performance of different implementations for common assistance problems. For describing those assistance problems, we introduce the Assistance Problem

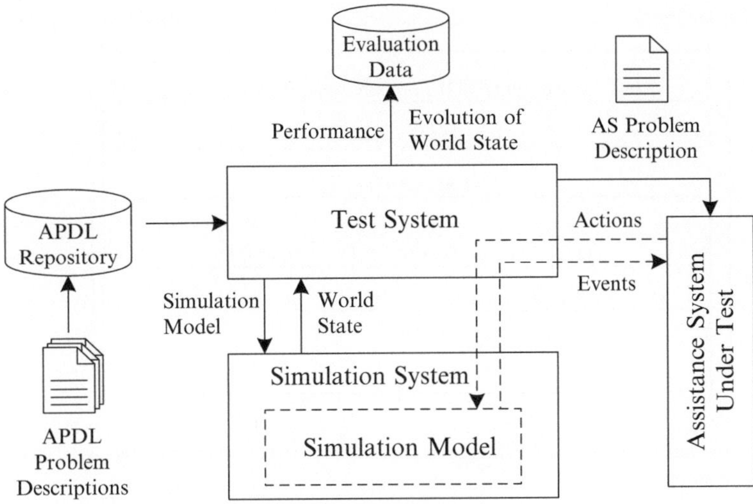

Fig. 1 System architecture. *Dashed lines* depict communication

Definition Language (APDL) that satisfies all language requirements identified in Sect. 3. This language is used to define problem descriptions for the test system and, at the same time, serves as input specification for the assistance system. A *problem description* refers to a machine-readable file defining an assistance problem.

Figure 1 depicts the overall architecture. The test system can load a certain APDL problem description from a repository. Afterwards, the test system extracts (a) an executable simulation model and (b) an AS problem description from the APDL file. The simulation model is passed to the simulation system and used to set up a simulation. It comprises an executable description for each object in the simulated smart environment. In contrast, the AS problem description hides information from the AS that is either of exclusive use for the simulation or "private" data. For instance the actual position of a user in a smart environment can be considered private, as an operating AS has to infer it from sensor readings. Usually, one does not want to use a fully executable model as the AS problem description.

Together with other test-specific parameters (such as time limits), the extracted problem description is sent to the AS under test. Afterwards, the test system passes the simulation model to the (third-party) simulation system and starts the simulation. Finally, the AS communicates with the simulation through our test system by sending action messages and receiving event messages. The communication is asynchronous. Both message types may be completely independent of the other: actions may not (immediately) generate an event, and not every event results in one action, or one event may result in a sequence of several actions.

Figure 2 shows a fragment of the communication between an AS and the simulation (model). There, the AS dims a lamp and afterwards receives an event that this lamp has changed its status. As the figure indicates, a test run comprises

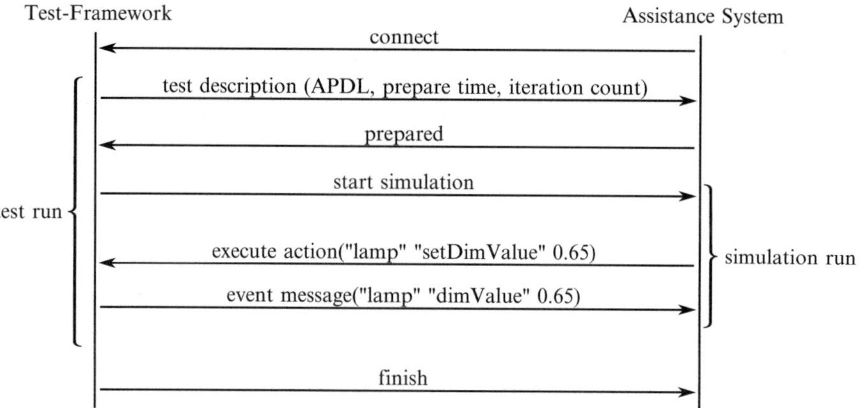

Fig. 2 Communication example. Messages from *right* to *left* are send from the assistance system to the simulation

several simulation runs. A simulation run is a single execution of the model (with a given parametrisation) on the simulation system, and resembles a single application of the AS to its problem domain. The overall procedure is called a test of an AS. One main difference between test runs and simulation runs is that the preparation time for parsing the APDL domain is only given once per test. The intention for this separation is two-fold. First, it is a convenient way for repeating randomised tests. Second, between two simulation runs the AS can and should transfer knowledge learned from previous runs (see also Sect. 5.2.1).

During each simulation run, our system repeatedly extracts the simulated world state from the simulation, i.e. the ground truth for the evaluation. This state as well as the actual performance of the AS are stored in the evaluation database and can be used for a systematic evaluation of the AS and a comparison of different ASs later on (see Sect. 5). The performance of an AS is computed based on the reward function specified within the APDL problem description.

4.1 A Simple Assistance Problem

The following simple multi-user assistance problem is used as an example through-out this section. Assume an abstract "stage lighting" domain, where a room is equipped with some lamps and a position sensor for the users. An AS can adjust each lamp's brightness and move them to any location.[4] The position sensor is able

---

[4]*Moving a lamp* is just an abstract concept. Assume the lamps are spot lights, where the target spot (or: angles between the lamp and the floor) can be set freely by an AS. We also abstract from any form of projection or distortion resulting in non-circular spot lights.

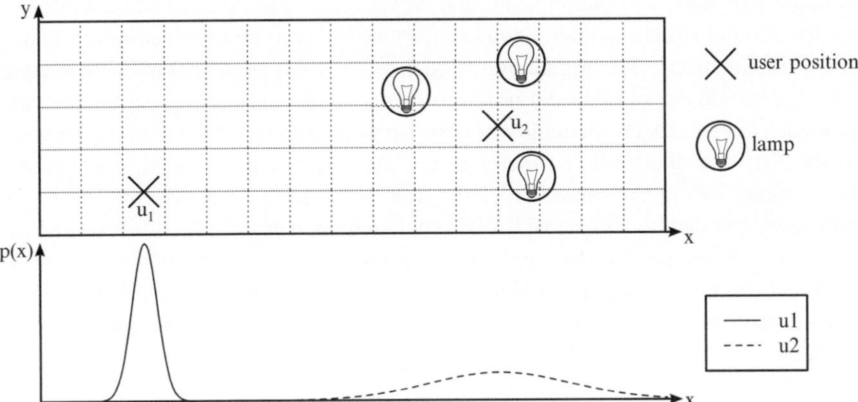

Fig. 3 Illustration of the sample assistance problem. Depicted is one room with three lamps and two users inside. At the *bottom*, the expected position sensor readings are shown

to estimate the users' positions. But its accuracy depends on the lighting conditions: The more light is cast to a person, the lower the accuracy of the sensor. The goal of the AS is to illuminate all users optimally by controlling the lamps, while the users move around freely. For this, an AS has to make a trade-off: illuminating the current user's position hinders estimating and thus illuminating the next user's position.

Figure 3 depicts how a setting may look like, where two users and three lamps are located in a room. For both users the probability density function of the position sensor is shown at the bottom. Note that the observations are more accurate for the user u_1 on the left, as all lamps are more closer to the user u_2 on the right. Over time, the users change their positions and the assistance system must adapt the lamps' positions.

4.2 *Assistance Problem Definition Language*

The *Assistance Problem Definition Language (APDL)* proposed here aims at describing assistance problems in smart environments, but is not restricted to them. The underlying concepts of the language are not completely new. Instead, APDL is designed as an extensible language. At its core, the language is defined to make as little assumptions on behaviour semantics as possible (while taking characteristic of smart environments into account). Extensions allow integrating different state-of-the-art behaviour descriptions.

We propose to define two different models representing an smart environment in APDL: (a) the *simulation model* for the simulation and the test system and (b) the *inference model* for the AS. In our test system, the state of the simulation model, which evolves during the simulation, represents the ground truth used for the evaluation of ASs. The inference model, on the other hand, is used as a description for the AS and should contain enough information for automated reasoning (using a reward function). The simulation model can exhibit more complex behaviour than one would or could model in the inference model. Separating simulation and inference model reflects the reality of ASs, because usually there exists no inference model that describes reality exactly and domain knowledge is always imperfect. Furthermore, using the same model for simulation and inference would imply to ultimately test the model against itself. Separate models instead facilitate testing the robustness of ASs and evaluating their performance, in particular when the inference model contains not only incomplete but also inaccurate information about reality. Notice, the input APDL file in Fig. 1 contains both the simulation and inference model.

Listing 2 Basic APDL example without action descriptions.

```
domain MovingLight {                        action setDim(float newDim);
  class User {                              action setPos(float[2] position);
    float[2] position :hidden;            }
    float totalLight :default = 0;       }
    action move() :private;
  }                                       reward :sum[User u: u.totalLight];
  class PositionSensor {
    User user :non-fluent;                instance MovingLight_2u_11
    float[2] position :event;               domain MovingLight {
    observes User.position;                 User user1 {position = [10.0, 0.0];},
  }                                             user2 {position = [0.0, 0.0];};
  class Lamp {                              PositionSensor ps_u1 {user = user1;},
    float dimValue :default = 0;             ps_u2 { user = user2; };
    float[2] position;                     Lamp lamp {position = [5.0, 5.0];};
                                          }
```

Because modelling an smart environment twice can be cumbersome, the core APDL language was designed to describe parts common to both the simulation and inference model. Extensions can be used to add more details for one of them independently. For instance, behaviour may be described in several ways. Here, we specify smart environments as discrete event systems for the simulation and use preconditions/effects for the inference model. If some new formalism proves to be more useful for AS inference in the future, it can be added to the language. This way, future ASs can be tested by our system, too.

4.2.1 The APDL Core

The core specification of APDL defines fundamental concepts such as object classes (devices and users), properties, and object instances. Smart environments—we think of—consist of many devices and sensors, but each belonging to one of just a few classes (e.g., lamps, projectors, or position sensors). It is then intuitive to use an object-oriented design. Encapsulating properties of objects (like devices in a smart room) in classes keeps objects independent of each other (which they usually are in reality). This leads to more modular and more easily maintainable models.

Listing 2 shows an incomplete formalisation of our simple example in APDL. As in PDDL and RDDL, the description is separated into problem domain and problem instance. A problem domain defines *object classes* and a problem instance defines for each class *object instances* including their initial state. The listing defines a domain with three classes representing the users, position sensors, and lamps of our running example. For simplicity, a separate position sensor for each user is assumed. The listing also contains an instance definition of the domain with two users and one lamp.

Every problem domain includes the definition of a reward function. This reward function plays a crucial role for the actual evaluation, as it is used by our test system as a performance measure indicating how "good" an assistance problem is solved. In Listing 2, each user has a property `totalLight` collecting how much the user is illuminated. The reward function is simply the sum of this property for all users.

When looking at assistance systems [7], problem definition languages [30, 31], and related formalisms [23], communication messages are divided into *actions* (or commands) and *events* (or observations). Actions change the state of an environment and may be executed by assistance systems. Changes in the environment may result in events sent to the AS. This distinction between the two different message types is also central to APDL and due to our asynchronous communication scheme.

The `observes` keyword defines the event messages an object is interested in. All properties of an object declared with `:event` define all events the object may generate. When an object receives an action message it executes the corresponding action as defined in the class. In the APDL core, actions are declared but not defined—the actual behaviour is defined for the simulation and inference model separately by using special extensions. Listing 2 defines three actions. Both actions of the lamp are "public" actions that may be executed by the AS. The `move` action of the user is annotated as private and the position property is hidden from the AS. Although not directly observable, an AS can recognise move actions by the position sensor.

4.2.2   Extension Mechanism

Extensions allow adding syntactic elements called *annotations*. Annotations can be added to many parts of an APDL file, such as object classes, properties, and actions. For instance, the predefined `:non-fluent` annotation indicates properties which never change their value.

Extensions are added by using the `:requires` annotation on problem domains as shown in Listing 3. There, e.g., the extension `PreconditionsEffects` has been added to the domain. This extension allows attaching the annotations `:precondition` and `:effect` to actions. We use preconditions and effects as a basis to define the inference model. Preconditions and effects are used as a fairly simple textual representation of state transitions. The semantics we propose are described in Sect. 4.4. The extension for the simulation model is described in Sect. 4.3.

Listing 3 shows the complete definition of the position sensor class as used in our example. Whenever a user's position (hidden from the AS) is updated its position sensor is informed, adds some noise to the real position, and updates its estimate (accessible by the AS). In this example the noise is modelled to be Gaussian (normally distributed), where the dim value of the lamp influences the covariance. Please remember, the actual simulation model may have a different definition of the `sensePosition` action, this is just an abstraction to be used by the AS.

Several extensions have been implemented [27]. For example probability distributions can be added. More domain knowledge can be added, e.g., with `TriggeredActions` that specifies *how* an object reacts to events. Extensions allow creating problem domains with different demands to ASs and varying complexity. For instance, the `Quantifiers` extensions enables the use of sum and product quantifiers, and the `Types` extensions is used to define available types for properties and hence controlling the state space complexity.

Listing 3 Definition of a position sensor. In this model, a position sensor can only track one user. The `TriggeredActions` extensions and the **observes** keyword allows to specify actions as response to events.

```
domain MovingLight                          action sensePosition(User u,
  :requires PreconditionsEffects,               float[2] userPos) :private
    PrivateActions, TriggeredActions,       :precondition { user == u }
    RandomFunctions(continuous.gauss) {     :effect { position =
  class PositionSensor {                        [gauss(userPos#0,
    User user :non-fluent;                        0.2 + u.currentLight),
    float[2] position :event;                   gauss(userPos#1,
                                                  0.15 + u.currentLight*1.4)];
    observes User.position :triggers       };
      sensePosition(:sender, :value);   }
                                       }
```

4.3   Integrating the Simulation Framework JAMES II

We exploit the simulation and experimentation framework JAMES II as simulation engine within our proposed test system. To specify simulation models and realise the communication between an AS under test and the simulation, we chose PepiDEVS, which is available as plugin for JAMES II. PepiDEVS distinguishes between atomic and coupled models. Atomic models represent components of the modelled system (here smart environments) and have a state, inputs, outputs, and a behaviour. Coupled models allow coupling these components and facilitate a hierarchical model design. The JAMES II -plugin requires PepiDEVS models to be implemented in a Java-based representation. For our prototype of the test system we define the simulation model directly by Java classes. Each class contains an atomic PepiDEVS model representing the associated APDL class in the simulation. The coupled model, which contains the "representatives" of all APDL classes and defines the couplings between those, is synthesized from the APDL problem description.

Using Java also allows us using other functionality provided by JAMES II, such as different random number generators or probability distributions. In addition to the Java extension, the test system provides an interface for controlling JAMES II and setting up a simulation. Instead of using JAMES II, other suitable simulation frameworks may be used by adding specific extensions and adapting the test system accordingly.

4.4 Formal Semantics

So far, we introduced and described our language APDL in a rather informal way. But we also developed a formal semantics [27], the concepts of which will be described here briefly. Using only the core language, an APDL domain defines:

- a set of classes $\mathscr{C}$ and set of properties $\mathscr{P}$, where the function $properties : \mathscr{C} \rightarrow 2^{\mathscr{P}}$ associates each class with a distinct set of properties,
- a set of all possible types $\mathscr{T}$ and a mapping $dom : \mathscr{P} \rightarrow \mathscr{T}$, which associates a type with each property,
- a set of all actions $\mathscr{A}$ and a function $actions : \mathscr{C} \rightarrow 2^{\mathscr{A}}$, declaring the actions for all classes,
- for every action $a \in \mathscr{A}$ with parameters $p_1, \ldots, p_n$, the function $dom(p_i)$ defining the parameter types and the tuple $signature(a) = (dom(p_1), \ldots, dom(p_n))$, and
- observation declarations $observations : \mathscr{C} \rightarrow 2^{\mathscr{P}}$, describing which class observes which properties.

A corresponding APDL instance defines a set $\mathscr{I}$ of object instances and a function $class : \mathscr{I} \rightarrow \mathscr{C}$ defining the class for each object instance, and the initial state s as a mapping $s : \mathscr{I} \times \mathscr{P} \rightarrow \mathscr{T}$ with $s(i, p) \in dom(p)$ for all instances $i \in \mathscr{I}$ and each property $p \in properties(class(i))$.

In particular we define the semantics of APDL by giving a Partially Observable Markov Decision Process (POMDP [23]) underlying the APDL domain. Informally, POMDPs define an optimal control problem given noisy observations and non-deterministic actions. They are a prominent formalism for solving real-world problems under uncertainty and have been used for assistance systems previously [17]. The APDL inference model may be used by ASs to reason about the environment and determine actions to execute. We therefore show how to transform the inference model to a POMDP. For this, now assume action semantics are defined by the preconditions/effects extension (Sect. 4.2.2). We also show the idea of how to track the current state of the world merely relying on observable events.

Formally, a POMDP is defined as a tuple (S, A, O, T, Ω, r). S is the set of possible states (e.g., user positions) and A is the set of possible actions (e.g., move). $T(s' \mid s, a)$ defines the probability distribution function (pdf) over successor states, given the current state s and executed action a. $\Omega(o \mid s', a)$ defines the pdf of observations $o \in O$ (position sensor readings) after execution of action a resulted in s'. The reward function r assigns a value to each transition and thus indicates good actions.

Our semantics for the inference model is based on an extended POMDP, which we define as a tuple $(P, P_E, S, A, A_S, T, \Omega, r)$. The set of possible states S, actions A, observations Ω, and the reward function r are as introduced above and are fully determined by APDL. Common POMDP inference techniques make it possible to track the current state of the world while relying on observable events only. Exploiting the information of devices and properties, we factor each state in a tuple $(v_1, \ldots, v_n)$ of independent values, each representing the value of a specific device's property. P is the flat set of all properties $i.p$ comprising the state space of each device i:

$$P = \{i.p \mid i \in \mathscr{I}, p \in properties(class(i))\}.$$

P_E is the set of all event properties. Similar to the properties, A is the set of all actions $i.a$ per device i, and A_S is the subset of all action executable by the assistance system (not marked as hidden or private). Given a problem instance containing two users, a position sensor for both, and one lamp of Listing 2, the set of properties is

$$P = \{u1.position, u1.totalLight, u2.position, u2.totalLight,$$

$$ps1.user, ps2.user, lamp.dimValue, lamp.position\}.$$

S is set of all possible states with $s(i.p) \in dom(p)$ for all $i.p \in P$. Both Ω and T are described by the preconditions and effects annotation for each action and can be automatically generated. T can be regarded as a factored distribution of all device properties' pdf with conditional dependencies (the current state depends on the last state). The same holds for the possible observations and Ω.

A challenge for an AS is to estimate the current state using the probability distribution function (pdf) $T(s' \mid s, a)$ of new states s' and the pdf $\Omega(o \mid s', a)$ of observations o, given an executed action a. An established and well-studied model for representing factored pdfs with conditional dependencies are Bayesian Networks (BNs). They have been extended to Dynamic Bayesian Networks (DBNs) for representing the evolution of such pdfs over time [24]. Hence, DBNs exactly resemble T, and BNs are suited to describe Ω. Conceptually, one DBN will be created for every action a of the inference model, where the nodes of the current (next) time slice are the properties of the current (next) state, respectively. The pdf $T(s' \mid s, a)$ of the POMDP is then simply the pdf defined by the DBN for action a, with the values of the current time slice taken from the current state s. If the precondition of an action a does not hold in a state s, then $T(s' \mid s, a)$ is undefined (or the only possible next state is a synthetic invalid state).[5] By analogy with T, Ω is comprised of one BN for every action a.

5 Evaluation

After the previous conceptual and theoretical presentation, this section now gives a practical evaluation of the system. First, it is shown how our test system can be used. Then, a working example of two problem domains with actual assistance systems is shown.

5.1 Usage of the Test System

The test system has been implemented in Java. It contains a parser for APDL and all listed extensions. A first version of the system can be obtained from http://mmis. informatik.uni-rostock.de/APDLtest. This web page contains all details with respect to downloading and running the system.

A single AS can be tested and evaluated with respect to a given assistance problem using our system as follows: The AS receives a problem description (denoted by AS Problem Description in Fig. 1) from our test system. This description is a processed version of the original APDL problem description, which contains both the simulation and inference model. After a predefined preparation time the simulation is started. Then, the AS interacts via our system with the simulation similarly as with a real environment, i.e. the AS can send action messages to control the simulated entities.

Our system extracts the simulated world state from the simulation. For each state, the reward function is computed and the result as well as the world state is stored in a special database. The AS can be evaluated based on this data and its behaviour over time. For instance, sudden drops at a specific time in the reward, in conjunction with the stored world state, can be used to debug problems. The history of executed actions may also be used to evaluate how quick an AS has acted in a given situation.

For comparing the performance of different ASs, they have to be evaluated separately as described above based on the same APDL problem description. However, due to factors external to the simulation, e.g., communication latency, reproducibility is only given to some extent. After all ASs are evaluated, their performance can be retrieved from the database and eventually compared. Similar to comparing different ASs, different configurations of a specific AS can be systematically compared.

---

[5] Handling invalid actions is beyond the scope here.

5.2 *Proof of Concept*

We modelled two basic assistance problems and implemented some simple ASs for demonstrating the applicability of our test system. The first problem domain for our evaluation is the running example used throughout Sect. 4. Designed for a more realistic setting, the second problem domain addresses adaptive and learning ASs.

For evaluation and debugging purposes, we developed (a) a "dummy" assistance system selecting purely random actions, (b) for each assistance problem a domain-specific AS, and (c) a graphical user interface to imitate assistance systems. Especially the latter is useful for manually "debugging" problem domains and their simulation model.

Domain-specific ASs are systems that are designed and tailored for a particular assistance problem in mind. This is how assistance is usually implemented for a given problem in a given environment. The random system serves as a general base-line system for comparison with the domain-specific systems. It can be used for many problem domains, because it has no hard-coded knowledge about any of

---

Listing 4 Complete definition of the Lamp class as used for the evaluation for the simple assistance systems.

```
domain MovingLight {                      action setDim(float newDim)
  class Lamp {                              :precondition
    float dimValue :default = 0.0;            { newDim >= 0, newDim <= 1 }
    float[2] position;                      :effect { dimValue = newDim; };
                                          action setPosition(float[2] newPos)
    // How often invalid dim values          :effect { position = newPos; };
    // were used in <setDim>.              }
    int invalidActions                  }
      :hidden :default = 0;             reward :sum[User u: u.totalLight]
                                          - 2 * :sum[Lamp l: l.invalidActions];
```

---

them. As soon as the random system receives the problem description from the test system, it extracts all actions it may execute. It has been designed as a purely reactive system: whenever an event from the simulated environment is received, with a certain probability, it executes a random action with a random parameter set (uniformly distributed). The following sections show the results of the random and domain-specific systems in two different problem domains.

5.2.1 The MovingLight Domain.

As described in Sect. 4.1, the MovingLight domain is only a synthetic example for demonstration purposes. The AS has the task of illuminating all users with a set of lamps, the position of which can be controlled by the AS. Noisy user observation, a limited area of light, and (in general) non-matching numbers of users and lamps are the main challenges. We used a problem instance with two users and three lamps.

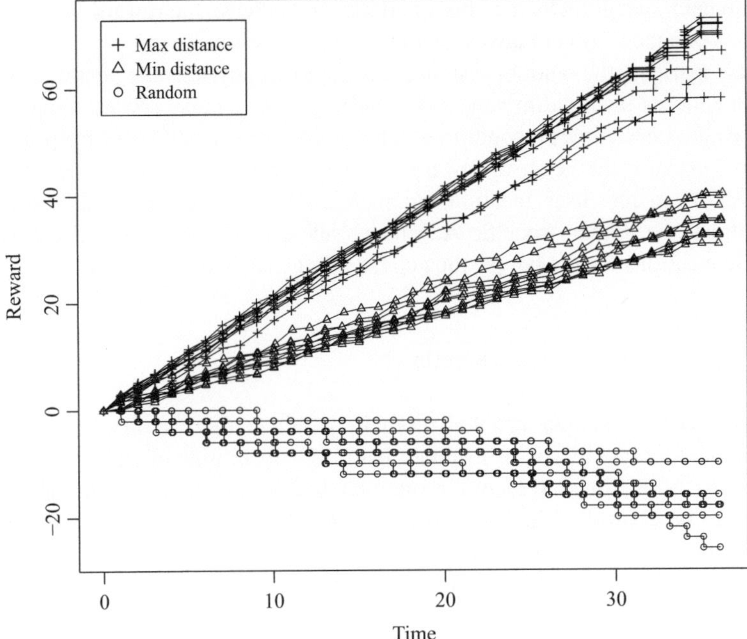

Fig. 4 Plot comparing the performance of domain-specific and random assistance systems for ten different simulation runs in the MovingLight domain

The specification of the lamp in the example (Listing 2) has been slightly extended (see Listing 4). The setDim precondition states that the new value must be in the range [0, 1]. Choosing an invalid value results in a negative reward.

For evaluation, we implemented two domain-specific ASs, with a similar strategy. Both systems maintain the observed positions of all users. Two central observations of the simple problem domain can be made and exploited. First, the light is just summed per user, so one user collecting all light gives the same reward as one light per user. It is therefore a good strategy to centre all lamps on the user with minimal noise on its position observation. Second, observation noise decreases when light shed onto the user also decreases. As a result, a naive strategy is: centre all lamps on the observed user position most distant from the current lamp positions.

The second domain-specific AS inverts this strategy (for demonstration purposes): The lamps are placed on the observed position nearest to the current position. It is expected that the "maximal distance" system is able to achieve a constantly higher reward than the "minimal distance" system. Figure 4 shows the rewards gained by the AS for ten different simulation runs, as measured by our test system.

Both domain-specific ASs are able to constantly increase the reward. As expected, the "max distance" system always achieves highest scores.[6]

In comparison, the random system is unable to set any sensible position. Instead, it often chooses invalid dim values and hence gets a negative reward—the random AS does not check any preconditions. The plot looks very different from the other systems, because the negative reward is -2 for invalid actions. As the lamps are far away from any user location, no increase in the reward can be perceived. In contrast, the domain-specific ASs never select invalid actions and can choose sensible positions for the lamps, thus constantly increase their reward.

5.2.2 The `ScheduleLearning` Domain

We now present another problem domain within a more realistic setting. The purpose of this problem is to evaluate learning capabilities of ASs, and is stated as follows: A meeting and lecture room contains two lamps, *stage* and *main*. Both lamps can be controlled independently by the AS, and are also controlled by the users of the room. The room is used according to a daily schedule. For instance, the first lecture lasts from 7:20 to 8:50, and a meeting starts at 10:10 and ends at 10:40. The lecture requires both lamps, and the meeting usually needs only *main*. The task

---

Listing 5 Model of the room for the `ScheduleLearning` domain. For the sake of brevity, some fields and all preconditions have been shortened.

```
domain ScheduleLearning {
class Room {
  int supportCount :hidden :default = 0;
  int manualControl :hidden :default = 0;

  // Each minute check required behaviour
  observes minute
    :triggers deactivateLamps()
    :triggers manualActivateLamps()
    :triggers activateLampsSupported();

  // At the end of each running session,
  // turn lamps off. (not rewarded)
  action deactivateLamps() :private
    :precondition { <session ended> }
    :effect {
      :action-call stage.setPower(false);
      :action-call main.setPower(false);
    } ;

  // Every minute of an active session,
  // check lamp configuration and switch
  // if necessary. This decreases reward
  action manualActivateLamps()
    :private
    :precondition {
      <lamps ill-configured>
      <active session>
    }
    :effect {
      :action-call stage.setPower(\ldots);
      :action-call main.setPower(\ldots);
      manualControl = manualControl + 1;
    } ;

  // At start of each session, check if
  // lamps were configured correctly.
  action activateLampsSupported()
    :private
    :precondition {
      <session has just started>,
      <lamps are configured correctly>
    }
    :effect {
      supportCount = supportCount + 1;
    } ;
}}
```

---

[6]This situation can of course change significantly when the domain and reward function changes. For instance when imposing a penalty on moving lamps, the maximising AS would get a much lower score and would most likely be outperformed by the minimal distance AS.

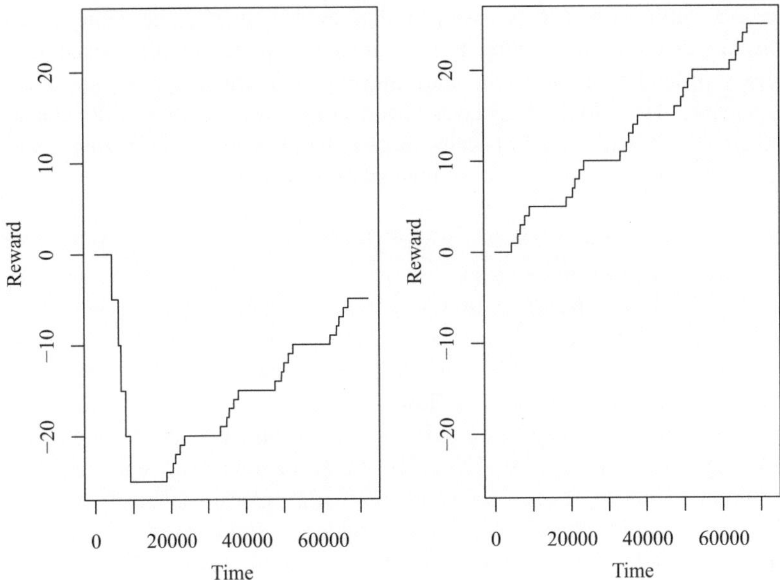

Fig. 5 Performance of two consecutive simulation runs (five days each) within one test run. The knowledge of the first run can be transferred to the second run

of the assistance system is to learn the schedule (which may change over time) and switch the lamps accordingly.

The APDL description of the central room model can be found in Listing 5. In this model, user behaviour is implicitly modelled as behaviour of the room object (a reasonable abstraction for this domain). If the assistance system fails to switch the lamps on for a scheduled lecture or meeting, the user (i.e. the room) will do so. In this case, a counter *manualControl* is increased. This can happen at any time during a session: if an AS changes the lighting conditions during a session, the room will intervene and switch back. If instead the system correctly adapts to the schedule, the counter *supportCount* is increased. During this evaluation we used the reward function $supportCount - 5 \cdot manualControl$ to punish switching lights during the course of a lecture or meeting.

One domain-specific AS has been developed, which serves as a demonstration for how to apply our test system to this domain. The assistance system is again very simple, but effective for this domain. During the very first day, it simply observes and saves all state changes of the lamps. This day is used as a prototype. Beginning with the second day, the lamp states are then simply reproduced.

Figure 5 depicts two simulation runs within one test run (cf. Sect. 4) of the AS. Each simulation run comprises five days. During the first run (shown on the left plot of Fig. 5), one can clearly see the drop in reward for each session (there are five sessions each day). After the system has learned the schedule, it is able to reproduce and gain reward. The second plot shows another simulation run. This time, the system could make use of its previous learned knowledge and was able to control the lamps right from the beginning of the simulation.

In this simple case, it is assumed that every day has the same schedule. Extensions of the problem domain could include a more complex schedule (e.g., a weekly instead of a daily schedule), and schedule changes from time to time. As a remark, the random AS receives rewards of approximately -1600, constantly decreasing over time. This is because it repeatedly switches the lamps randomly during a session, resulting in a lot of manual user interactions.

Conclusions and Future Work

An assistance system (AS) should support users of smart environments proactively, by executing useful and "intelligent" actions automatically. Rigorous testing is necessary to ensure that such assistance effectively and safely works under various conditions. However, testing assistance systems in real environments can be expensive or even not possible at all and testing conditions are hard to reproduce. The latter makes an "objective" comparison of different assistance systems based on real-world experiments barely possible. Therefore, we developed an assistance problem definition language (APDL) and a test system that allows testing ASs based on simulation. An AS can be evaluated by using reward data generated by the test system. A systematic evaluation using the same problem description and simulated environment enables a direct comparison of different approaches, an aspect missing in most of the systems realised so far.

APDL has been designed as an extensible language to describe assistance problems application-independently. The language provides means for a formal specification of assistance problems at an abstract level, including an explicitly defined reward function. An inference model is used by the AS to track/infer the state of the environment and to estimate good actions to support the user. The essential reward function is used to define the quality of the provided assistance formally and thus facilitates a comparison of the performance of different systems in the first place. However, defining suitable reward functions is a challenge that has to be tackled by the developers of concrete smart environments, as "good" assistance depends on the particular application scenario.

In addition to the test system itself, some (rather simple) assistance systems have been developed and used to provide a proof of concept for the test system. Nonetheless, some problems remain to be addressed and to be solved. For further evaluation, many different use cases from various application domains should be specified in APDL. So far, we specified different assistance problems related to our research domain, i.e. smart meeting rooms. But as argued above, our language itself is not restricted to these types of smart environments and can also be used to specify other assistance problems from domains, such as smart homes and smart cities. We also discussed the formal

(continued)

semantics underlying our language, by mapping it to POMDPs. As a next step, we will investigate the expressive power of various extensions (and their combination) in detail. In particular, we will try to identify extensions that enable an optimal solution of the assistance by solving the underlying POMDP directly.

One of the original motivations for this research was the idea of an *universal assistance system*, which is a system that can be deployed in different environments and solves the resulting specific assistance problems by simply being provided with an appropriate problem description. Such application-independent assistants may evolve into personal assistants accompanying their users in different settings while supporting the users pro-actively. Universal ASs will benefit largely from advances in AI research. Even though there exists no such universal assistance system yet, we believe that the proposed language with its clearly defined formal semantics is a necessary first step towards the development of such an assistance systems, and that our test system can be used to evaluate those assistance systems systematically.

Acknowledgements This research is supported by the German Research Foundation (DFG) within the research training group GRK 1424 MuSAMA (Multimodal Smart Appliance Ensembles for Mobile Applications).

References

1. Bader, S., Dyrba, M.: Goalaviour-based control of heterogeneous and distributed smart environments. In: Proceedings of the 7th International Conference on Intelligent Environments, pp. 142–148. IEEE Computer Society (2011)
2. Bayha, A., Grüneis, F., Schätz, B.: Model-based software in-the-loop-test of autonomous systems. In: Wainer, G.A., Mosterman, P. (eds.) Proceedings of the 2012 Symposium on Theory of Modeling and Simulation—DEVS Integrative M&S Symposium. Society for Computer Simulation International, Article No. 30 (2012)
3. Brooks, R.A.: The intelligent room project. In: Proceedings of Second International Conference on Cognitive Technology Humanizing the Information Age, pp. 271–278 (1997)
4. Chin, J., Callaghan, V., Clarke, G.: An end user tool for customising personal spaces in ubiquitous environments. In: Proceedings of the Third international conference on Ubiquitous Intelligence and Computing, UIC'06, pp. 1080–1089. Springer, Berlin/Heidelberg (2006)
5. Cook, D.J., Das, S.K.: Smart Environments. Wiley, Hoboken (2005)
6. Cook, D.J., Huber, M., Gopalratnam, K., Youngblood, M.: Learning to control a smart home environment. In: Riedl, J., Hill, R. (eds.) Innovative Applications of Artificial Intelligence. The AAAI Press, Menlo Park, (2003)
7. Cook, D.J., Youngblood, M., Das, S.K.: A multi-agent approach to controlling a smart environment. In: Designing Smart Homes. Lecture Notes in Computer Science, vol. 4008, pp. 165–182. Springer, Berlin/Heidelberg (2006)
8. Dooley, J., Callaghan, H.H.V., Gardner, M., Ghanbaria, M., AlGhazzawi, D.: The intelligent classroom: beyond four walls. In: Proceedings of the Intelligent Campus Workshop (IC'11) held at the 7th IEEE Intelligent Environments Conference (IE'11), Nottingham (2011)

9. Genesereth, M., Love, N.: General game playing: overview of the AAAI competition. AI Mag. **26**, 62–72 (2005)
10. Ghallab, M., Howe, A., Knoblock, C., Mcdermott, D., Ram, A., Veloso, M., Weld, D., Wilkins, D.: PDDL—the planning domain definition language. Technical Report CVC TR-98-003/DCS TR-1165, Yale Center for Computational Vision and Control (1998)
11. Gierke, M., Himmelspach, J., Röhl, M., Uhrmacher, A.M.: Modeling and simulation of tests for agents. In: Fischer, K., Timm, I.J., André, E., Zhong, N. (eds.) Multiagent System Technologies. Lecture Notes in Computer Science, vol. 4196, pp. 49–60. Springer (2006)
12. Giersich, M., Heider, T., Kirste, T.: AI methods for smart environments: a case study on team assistance in smart meeting rooms. In: Proceedings of Scientific Workshop 1: Artificial Intelligence Methods for Ambient Intelligence on European Conference on Ambient Intelligence (AmI-07), Darmstadt, Germany (2007)
13. Helal, S., Mann, W., El-Zabadani, H., King, J., Kaddoura, Y., Jansen, E.: The gator tech smart house: a programmable pervasive space. Computer **38**(3), 50–60 (2005)
14. Helal, S., Lee, J.W., Hossain, S., Kim, E., Hagras, H., Cook, D.J.: Persim—simulator for human activities in pervasive spaces. In: Proceedings of the Seventh International Conference on Intelligent Environments, pp. 192–199. IEEE Computer Society (2011)
15. Himmelspach, J.: Konzeption, Realisierung und Verwendung eines allgemeinen Modellierungs-, Simulations- und Experimentiersystems—Entwicklung und Evaluation effizienter Simulationsalgorithmen, vol. 4, 1st edn. Sierke (2007) in German
16. Himmelspach, J., Uhrmacher, A.M.: Plug'n simulate. In: Proceddings of the 40th Annual Simulation Symposium, pp. 137–143. IEEE Computer Society (2007)
17. Hoey, J., Poupart, P., von Bertoldi, A., Craig, T., Boutilier, C., Mihailidis, A.: Automated handwashing assistance for persons with dementia using video and a partially observable markov decision process. Comput. Vis. Image Underst. **114**(5), 503–519 (2010)
18. Huebscher, M.C., McCann, J.A.: Simulation model for self-adaptive applications in pervasive computing. In: Proceedings of the Database and Expert Systems Applications, pp. 694–698 (2004)
19. Kientz, J.A., Patel, S.N., Jones, B., Price, E., Mynatt, E.D., Abowd, G.D.: The Georgia Tech Aware Home. In: CHI '08 Extended Abstracts on Human Factors in Computing Systems, pp. 3675–3680. ACM, New York (2008)
20. Krüger, F., Steiniger, A., Bader, S., Kirste, T.: Evaluating the robustness of activity recognition using computational causal behavior models. In: Dey, A.K., Chu, H.-H., Hayes, G. (eds.) Proceedings of the 2012 ACM Conference on Ubiquitous Computing, pp. 1066–1074. ACM Press, New York (2012)
21. Hambling, B., Morgan, P., Samaroo, A., Thompson, G., Williams, P.: Software Testing: An ISTQB-ISEB Foundation Guide, revised edn. British Informatics Society Ltd., Swindon (2010)
22. Mozer, M.C.: Lessons from an adaptive house. In: Cook, D.J., Das, R. (eds.) Smart Environments: Technologies, Protocols, and Applications, pp. 273–294. Wiley, Hoboken (2005)
23. Murphy, K.P.: A survey of POMDP solution techniques. Technical report, Department of Computer Science, University of British Columbia (2000)
24. Murphy, K.P.: Dynamic bayesian networks: representation, inference and learning. Ph.D. thesis, University of California, Berkeley (2002)
25. Nakashima, H., Aghajan, H., Augusto, J.C. (eds.): Handbook of Ambient Intelligence and Smart Environments. Springer, New York (2010)
26. Nishikawa, H., Yamamoto, S., Tamai, M., Nishigaki, K., Kitani, T., Shibata, N., Yasumoto, K., Ito, M.: UbiREAL: Realistic smartspace simulator for systematic testing. In: Dourish, P., Friday, A. (eds.) Proceedings of the 8th international conference on Ubiquitous Computing. Lecture Notes in Computer Science, vol. 4206, pp. 459–476. Springer (2006)
27. Nyolt, M.: Concept and realisation of a framework for testing general assistance systems. Master's thesis, University of Rostock (2012)
28. Pfeffer, A.: IBAL: a probabilistic rational programming language. In: Proceedings of the 17th International Joint Conference on Artificial Intelligence, vol. 1, pp. 733–740. Morgan Kaufmann Publishers Inc., San Francisco (2001)

29. Poslad, S.: Ubiquitous Computing: Smart Devices, Environments and Interactions.Wiley, Chichester (2009)
30. Proetzsch, M., Zimmermann, F., Eschbach, R., Kloos, J., Berns, K.: A systematic testing approach for autonomous mobile robots using domain-specific languages. In: KI 2010: Advances in Artificial Intelligence. Lecture Notes in Computer Science, vol. 6359, pp. 317–324. Springer (2010)
31. Sanner, S.: Relational dynamic influence diagram language (RDDL): language description. http://users.cecs.anu.edu.au/~ssanner/IPPC_2011/RDDL.pdf (2010)
32. Sato, T.: Logic-based probabilistic modeling. In: Ono, H., Kanazawa, M., de Queiroz, R. (eds.) Logic, Language, Information and Computation. Lecture Notes in Computer Science, vol. 5514, pp. 61–71. Springer, Berlin/Heidelberg (2009)
33. Uhrmacher, A.M., Röhl, M., Himmelspach, J.: Unpaced and paced simulation for testing agents. In: Verbraeck, A., Hlupic, V. (eds.) Proceedings of the 15th European Simulation Symposium, pp. 71–80. SCS European Publishing House (2003)

A Relaxation Strategy with Fuzzy Constraints for Supplier Selection in a Power Market

Dien Tuan Le, Minjie Zhang, and Fenghui Ren

Abstract A power market is a special kind of e-market. In a power market, all trading processes are related to three parties: buyers, suppliers and brokers. A broker acts as middlemen between buyers and suppliers in a trading process. In a power market, how to select a potential supplier for a buyer through a broker based on the buyer's requirements is a challenging research problem. This paper proposes relaxation strategy with fuzzy constraints for supplier selection. The strategy includes three components, i.e., a supplier selection, a fuzzy constraint relaxation, and a decision making. The major contributions of this paper are that (1) the trading process between buyers and suppliers through brokers is modeled by using fuzzy constraints through the consideration of multiple attributes of the buyer's requirements as well as potential power suppliers; and (2) a buyer can utilize a relaxation with fuzzy constraints to change its requirements in difficult situations when a broker cannot find any supplier to satisfy a buyer's requirements. Experimental results show that our approach is successfully applied in a simulated power market.

Keywords Broker • E-market • Power market • Trading agent

1 Introduction

E-markets are virtual marketplaces where buyers and suppliers meet to exchange information about price, products and service offerings, and to negotiate and carry out business transactions [1]. A power market is a kind of e-market with specifications. In a power market, a buyer cannot reach a contract with a supplier directly and any supplier selection must go through a broker. Thus, in a power market involving three parties, it is hard to apply game theory-based negotiation approaches, which have been successfully applied in many e-markets, to reach an agreement in trading processes. In addition, because of electricity energy's

D.T. Le (✉) • M. Zhang • F. Ren
University of Wollongong, Wollongong, NSW, Australia
e-mail: dtl844@uowmail.edu.au; minjie@uow.edu.au; fren@uow.edu.au

© Springer Japan 2015

Q. Bai et al. (eds.), *Smart Modeling and Simulation for Complex Systems*, Studies in Computational Intelligence 564, DOI 10.1007/978-4-431-55209-3_6

constraints from policies of companies or organizations, it is difficult for buyers and suppliers to use automated negotiation approaches [3, 8, 9] or auction mechanisms [7, 12]. Due to the involvement of third party (brokers) [5], a broker in a power market acts as middlemen between buyers and suppliers to help a buyer identify a potential supplier in order to reach an electricity supply contract.

The study of brokers, acting as the third party of a trading process in e-markets, has been examined by many researchers in recent years. Sarma et al. [11] provided an efficient algorithm to compute the equilibrium for a related game of price in a trading process between buyers and sellers through brokers. Blume et al. [2] studied the trading process in general e-markets between buyers and sellers through a layer of intermediaries. Rubenstein et al. [10] proposed a market model with three types of agents, i.e., sellers, buyers, and middlemen, and analyzed steady state conditions in such markets. Gale et al. [4] analyzed a network model of exchange, in which trading is intermediated. Due to the special features of power markets, these approaches are facing two main challenges in power markets: (1) every trading process between buyers and suppliers in a power market is related to a broker, so a broker plays an important role in the supplier selection; and (2) due to policies of organizations or companies, suppliers cannot make any concession to a buyer.

To address the above challenges, we develop a relaxation strategy with fuzzy constraints to select a potential power supplier to agree on an electricity supply contract through a broker. In particular, our approach uses prioritized fuzzy constraints to present trade-offs between the different possible values of attributes and to indicate how relaxations should be made when they are necessary. The major contributions of this paper are that (1) the trading process between buyers and suppliers through brokers is successfully addressed based on fuzzy constraints for multiple attributes of buyer's requirements as well as potential power suppliers; and (2) a buyer utilizes a relaxation strategy with fuzzy constraints to change its requirements when a broker cannot find any supplier to meet a buyer's requirements. Experimental results demonstrate the good performance of the proposed approach in terms of satisfaction of buyer's requirements.

The rest of this paper is organized as follows. The problem description and definitions are presented in Sect. 2. A relaxation strategy with fuzzy constraints to select a potential power supplier is introduced in Sect. 3. An experiment is presented in Sect. 4. Section 5 concludes this paper and points to our future work.

2 Problem Description and Definitions

In general, there are three parties involving in a power market, i.e., suppliers, buyers and brokers. One objective of the trading process between a buyer and suppliers through a broker is to select a potential supplier for the buyer. To achieve the objective, the three parties in the trading process have to follow certain rules. (1) multiple suppliers provide their electricity to the power market; (2) due to policies of companies or organizations, suppliers cannot make any concession to buyers;

and (3) a buyer's requirements must be satisfied by suppliers. Before elaborating the details of the approach, it is necessary to define the scope of this research and provide some necessary definitions.

A buyer agent is considered as an electricity consumer who would like to find a potential power supplier and agree on a contract.

Definition 1. A buyer agent B_i is defined as a 4-tuple $B_i =< ID, EER, \alpha, \lambda >$, where ID is the buyer agent's identification, EER indicates the electricity energy request (see Definition 2), α is the acceptability threshold of B_i, and λ is the concession threshold of B_i.

Definition 2. An electricity energy request is represented by EER and is defined by the following format.

$$EER = \begin{pmatrix} A_1 & A_2 & \dots & A_n \\ C_1 & C_2 & \dots & C_n \\ W_1 & W_2 & \dots & W_n \end{pmatrix}, \tag{1}$$

where A_i indicates the ith attribute name, C_i is the constraint value of A_i and W_i is the priority value of A_i, $1 \leq W_i \leq n$. $W_i=1$ indicates the lowest priority and $W_i = n$ indicates the highest priority.

A supplier agent is considered as a company or an organization and its responsibility is to sell electricity to buyer agents in a power market.

Definition 3. A supplier agent S_i is defined as a 3-tuple $S_i =< ID, ER, BO >$, where ID is the identification of S_i, ER indicates an electricity resource (see Definition 4) provided by S_i, and BO is a bonus value which a supplier agent may use to attract a buyer agent to purchase electricity.

The electricity resource is provided by a supplier agent. Due to company or organization policies, an offer from a supplier agent contains some constraints to electricity resource such as constraints of price and time.

Definition 4. An electricity resource provided by a supplier agent S_i is presented by ER and is defined by the following format.

$$ER = \begin{pmatrix} A_1 & A_2 & \dots & A_n \\ C_1 & C_2 & \dots & C_n \end{pmatrix}, \tag{2}$$

where A_i is the ith attribute name and C_i is the constraint value of A_i provided by S_i.

A broker agent acts as the third party in a trading process between a buyer agent and supplier agents. A broker agent's responsibility is to select the most suitable supplier agent to meet a buyer agent's requirements.

Definition 5. A broker agent BR_i is defined as $BR_i =< S, B_i, r >$, where S is a set of supplier agents, which sell electricity to a buyer agent B_i, r is a reward that BR_i can get from a supplier agent.

3   A Relaxation Strategy with Fuzzy Constraints for Supplier Selection

Our relaxation strategy includes the three main components: (1) the supplier selection; (2) the relaxation with fuzzy constraints; and (3) the decision making. In this section, the principle of the whole trading process is introduced in Sect. 3.1. Then the three main components are presented in details in three subsections, respectively.

3.1   The Principle of the Whole Trading Process

3.1.1   Background

A trading process between a buyer agent and supplier agents is conducted through a broker agent to achieve an agreement by using certain strategies. In our strategy, a buyer agent utilizes the relaxation with fuzzy constraints to change its requirement in difficult situations. The broker agent relies on a reward from supplier agents to select the most suitable supplier agent for the buyer agent. Supplier agents use a bonus policy to attract the buyer agent to purchase their electricity. The principle of the whole trading process between a buyer agent and supplier agents through a broker agent in our approach is presented in Fig. 1.

Step 1:   The buyer agent selects a constraint of attributes with the highest priority from its requirements and sends the constraint to the broker agent. Based on the buyer agent's constraint, the broker agent searches for supplier agents. If the broker agent cannot find any supplier agent, the broker agent checks whether the buyer agent's constraints can be relaxed. If the relaxation is not applied, the trading process is terminated. Otherwise, the buyer agent selects a relaxed constraint and sends it to the broker agent again. This procedure will be repeated until the broker agent finds supplier agents to satisfy the buyer agent's constraints or the trading is terminated.

Step 2:   Once the broker agent finds suitable supplier agents, it will select the most suitable supplier agent based on the suitable supplier agents' rewards and send the selected supplier agent to the buyer agent. The buyer agent checks whether the most suitable supplier agent satisfies the buyer agent's other constraints. If there are more constraints, the buyer agent selects the next highest priority constraints and sends it to the broker agent again, and the process goes to step 1. Otherwise, the buyer agent evaluates whether the most suitable supplier agent is acceptable.

Step 3:   If the buyer agent accepts the most suitable supplier agent, the trading process makes a deal. Otherwise, the buyer agent requires the broker agent to check whether the most suitable supplier agent offers a bonus. If the most suitable

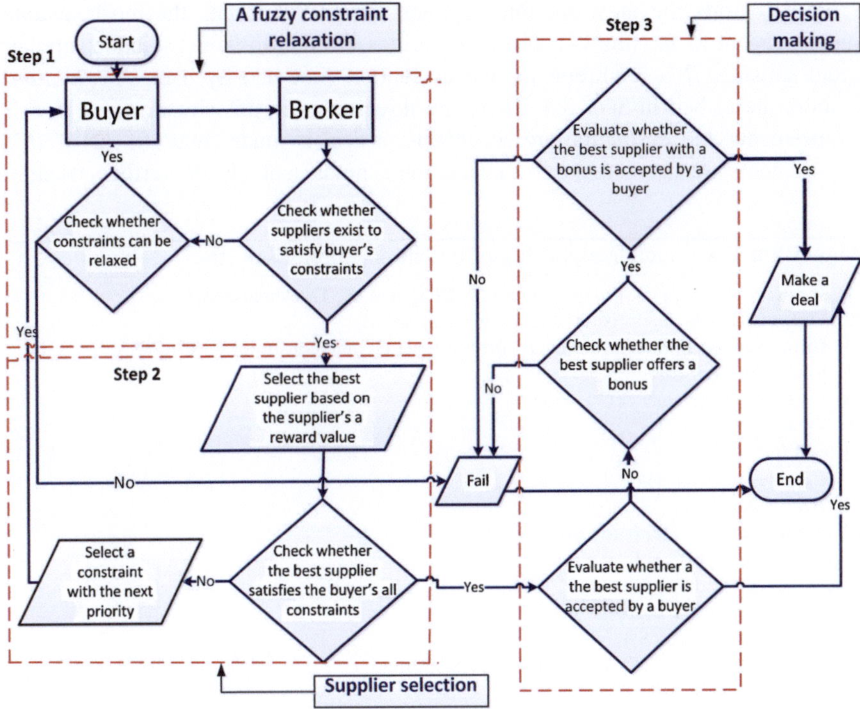

Fig. 1 Diagram of the principle

supplier agent does not offer a bonus, the trading process between the buyer agent and the broker agent is terminated. Otherwise, the buyer agent evaluates the most suitable supplier agent with a bonus again to make a decision.

3.1.2 Formal Description

A formal representation of the process of potential supplier selection is described by Algorithm 1.

The algorithm shows all cases of the trading process between a buyer agent B_i and a set of supplier agents $\mathbf{S}$ through a broker agent BR_i based on the B_i's requirements, an acceptability threshold and a concession threshold (line 1). The output of the algorithm can be either *'deal'* or *'fail'* (line 2).

First, B_i uses its acceptability threshold to determine each constraint value of an attribute in EER (lines 5, 6). Then B_i selects a constraint of an attribute in EER with the highest priority and sends it to BR_i (lines 8, 9). BR_i finds the most suitable supplier agent to satisfy B_i's requirements (line 12) by using *'find'* function described in Sect. 3.2. The results from BR_i are presented as follows.

If BR_i finds the most suitable supplier agent, BR_i sends the most suitable supplier agent to B_i (line 14). Then, B_i verifies whether the most suitable supplier agent satisfies B_i's requirements and evaluation (line 15) by using *'evaluation'* function described in Sect. 3.4. There are three cases in this situation. (1) if B_i's requirements and evaluation are acceptable, a deal is made (line 16). (2) if B_i's requirements are satisfied but B_i's evaluation is not acceptable, B_i verifies whether

---

Algorithm 1 A principle algorithm of potential supplier selection

---

1: **Input:** $\mathbf{S} = \{S_i \mid i = \overline{1, n}\}$, $B_i = <ID, EER, \alpha, \lambda>$. Threshold $\alpha, \lambda \in [0, 1]$
2: **Output:** Return the decision of making a deal or fail
3: **Initialization:** Initialize submitted-constraint-set $\mathbf{C}^*$ and constraint set $\mathbf{C}$ to $\varnothing$
4: **for** $\forall$ i in EER **do**
5:     $C_i \leftarrow$ determine($f(C_i) \geq \alpha$)
6:     $\mathbf{C} \leftarrow \mathbf{C} \cup \{C_i\}$
7: **end for**
8: $C_{new} \leftarrow$ argmax$_{\mathbf{C}}(W_i)$
9: $BR_i \leftarrow$ send(C_{new})
10: **while** $\neg stopCriterion()$ **do**
11:     $\mathbf{C}^* = \mathbf{C}^* \cup \{C_{new}\}$
12:     $S' \leftarrow$ find($\mathbf{C}^*$,$\mathbf{S}$)
13:     **if** $S' \neq Null$ **then**
14:         $B_i \leftarrow$ send(S')
15:         **if** check($\mathbf{C}^*$,S') and evaluate($\mathbf{C}^*$,S',0) **then**
16:             success()
17:         **else**
18:             **if** check($\mathbf{C}^*$,S') and $\neg$ evaluate($\mathbf{C}^*$,S',0) **then**
19:                 **if** $B_i \leftarrow$ offer-bonus(S') and evaluate($\mathbf{C}^*$,S',BO) **then**
20:                     success()
21:                 **else**
22:                     fail()
23:                 **end if**
24:             **else**
25:                 $C_{new} \leftarrow$ argmax$_{\mathbf{C} \backslash C_{new}}(W_i)$
26:                 $BR_i \leftarrow$ send(C_{new})
27:             **end if**
28:         **end if**
29:     **else**
30:         **if** $B_i \leftarrow$ relax($\mathbf{C}^*$) **then**
31:             $B_i \leftarrow$ update(EER)
32:             Go to line 4
33:         **else**
34:             fail()
35:         **end if**
36:     **end if**
37: **end while**

---

the most suitable supplier agent offers a bonus. If the most suitable supplier agent offers the bonus and B_i's evaluation with a bonus is acceptable, the trading process between B_i and BR_i makes a deal. (lines 18–20). Otherwise, the trading process between B_i and BR_i is terminated (line 22). (3) if B_i's requirements are not satisfied, B_i selects a constraint of attributes with the next highest priority in the EER and sends it to BR_i (lines 25–26). Thus, BR_i has to find suitable suppliers again with the new constraints.

If BR_i does not find any suitable supplier agent, which satisfies B_i's requirements, B_i has to relax its requirements (line 30) by using *'relaxation'* function described in Sect. 3.3. In particular, if a constraint of an attribute is relaxed by B_i, B_i has to update its EER and the algorithm runs again with the updated EER (lines 31, 32). Otherwise, the trading process is terminated (line 34).

The three major components of the proposed approach are introduced in detail in the following three subsections, respectively.

3.2 Supplier Selection

When a broker agent receives the buyer agent's requirements, the broker agent starts to find the most suitable supplier agent for a buyer agent. The *'find'* function, displayed in line 12 of Algorithm 1, is shown in Algorithm 2 as follows.

Algorithm 2 find($\mathbf{C}^*, \mathbf{S}$)

```
 1: Input: S = {S_i | i = 1,n}, a set of constraints C*
 2: Output: return the most suitable supplier or null
 3: for each S_i in S do
 4:      add← true
 5:      for each C_i in C* do
 6:          if f(S_i.C) ≤ f(C_i) then add← false
 7:          end if
 8:      end for
 9:      if add=true then
10:          SS ← SS ∪ {S_i}
11:      end if
12:      if SS is not ∅ then
13:          return argmax_{S_i ∈ SS}(S_i.r)
14:      else
15:          return Null;
16:      end if
17: end for
```

Algorithm 2 shows how to select the most suitable supplier agent based on a set of suppliers $\mathbf{S}$, a set of constraints called $\mathbf{C}^*$ which has been submitted to BR_i during the selection stage (line 1). The output of the algorithm can be either *'the most suitable supplier'* or *'null'* (line 2). First, BR_i selects a suitable supplier set, which satisfies B_i's requirements (lines 3–10). Then, the most suitable supplier agent is selected from the supplier set based on a maximal reward value from supplier agents. If BR_i finds the most suitable supplier agent, BR_i sends it to B_i (line 13). Otherwise, BR_i cannot find any supplier agent which satisfies B_i's requirements (line 15).

3.3 A Fuzzy Constraint Relaxation

If BR_i cannot find any S_i, which satisfies B_i's requirements, BR_i requests B_i to consider relaxing its requirements. The *'relaxation'* function, displayed in line 26 of Algorithm 1, will be activated. The relaxation function is shown in Algorithm 3.

The algorithm shows how to carry out the relaxation based on a set of constraints called $\mathbf{C}^*$, which has been submitted to BR_i so far and the concession threshold λ (line 1). The output of the algorithm can be either *'a selected constraint for the relaxation'* or *'false of the relaxation'* (line 2). The algorithm proceeds as follows.

Algorithm 3 relax(C^*)

1: **Input:** a set of constraints $\mathbf{C}^*$, the concession threshold λ
2: **Output:** return a selected constraint for a relaxation or null
3: k $\leftarrow$ argmax$_{EER}(W_i)$
4: l $\leftarrow$ inf
5: $C^k \leftarrow$ Null
6: **for Each** C in $\mathbf{C}^*$ **do**
7:     **if** $f(C^R) \geq \lambda$ **then**
8:         d $\leftarrow f(C_i) - f(C^R)$
9:         p $\leftarrow W_i / k$
10:     **end if**
11:     **if** d*p<l **then**
12:         l $\leftarrow d * p$
13:         $C^k \leftarrow C_i$
14:     **end if**
15: **end for**
16: return C^k;

After determining the highest priority in EER (line 3), B_i checks whether each constraint of an attribute in C^* is satisfied for the relaxation. This means that B_i determines the degrees of satisfaction for the relaxation of each constraint. When constraint C of an attribute is decreased to the next highest satisfaction degree, the decreased constraint is named C^R. If a satisfaction degree of a relaxed constraint $f(C^R)$ is less than its concession threshold λ, the relaxation of the constraint is not permitted. Otherwise, the constraint is considered for a relaxation. The process of the relaxation is illustrated as follows. First, B_i calculates a decreased satisfaction degree (line 8) and a relative priority degree (line 9) for each constraint of an attribute in C^*. Then, a lost benefit value for each constraint after relaxation is calculated from a decreased satisfaction degree and a relative priority degree. Based on a lost benefit value for each relaxed constraint, B_i selects a constraint for a relaxation with the smallest lost benefit to B_i (lines 12, 13).

3.4 Decision Making

The *'evaluation'* function, displayed in line 15 of Algorithm 1, is shown in Algorithm 4. Algorithm 4 presents how to evaluate the most suitable supplier agent based on B_i's updated EER, the acceptability threshold, the bonus from the most suitable supplier agent (line 1). The output of the algorithm can be either *'acceptability'* or *'unacceptability'* (line 2). The algorithm proceeds as follows.

Algorithm 4 evaluate($\mathbf{C}^*$,S',BO)

1: **Input:** constraint set $\mathbf{C}^*$, the most suitable supplier S', and a bonus BO
2: **Output:** return true if satisfaction or false if unsatisfaction
3: k $\leftarrow$ argmax$_{EER}(W_i)$
4: $\delta \leftarrow inf$
5: **for Each** C_i in $\mathbf{C}^*$ **do**
6:     p $\leftarrow W_i/k$
7:     $t \leftarrow (f(C_i) - 1) * p + 1$
8:     **if** $t < \delta$ **then**
9:         $\delta \leftarrow t$
10:    **end if**
11: **end for**
12: $\Delta_{ap} \leftarrow \Delta(\alpha, \gamma, \delta)$
13: return $(\Delta_{ap} > \alpha)$;

B_i calculates an acceptability degree called Δ_{ap} to compare to α. The acceptability degree is related to three parameters δ, γ, and α [6]. Parameter $\delta \in [0, 1]$ is called the overall satisfaction degree and is calculated from B_i's updated EER. To calculate δ value, we calculate corresponding suitable degree t_i for each constraint C_i (lines 6, 7). Then, δ value is $min\{t_i\}$ (line 9). Parameter $\gamma \in [0, 1]$ is the satisfactory degree of a bonus from S'. Parameter α is the acceptability threshold of B_i. Based on δ, γ, and α [6], Δ_{ap} is calculated from Eq. 3 as follows (line 12).

$$\Delta_{ap}(\alpha, \gamma, \delta) = \frac{(1 - \alpha)\delta((1 - \alpha)\gamma + \alpha)}{(1 - \alpha)\delta((1 - \alpha)\gamma + \alpha) + \alpha(1 - \delta)(1 - ((1 - \alpha)\gamma + \alpha))} \tag{3}$$

If Δ_{ap} is more than α, the most suitable supplier agent is acceptable. Otherwise, the most suitable supplier agent is unacceptable (line 13).

4   Experiment

In this section, we illustrate our experimental results on relaxation strategy with fuzzy constraints for supplier selection in the power market. Section 4.1 introduces the experimental setting. Section 4.2 demonstrates the experimental results.

4.1   Experiment Setting

The experiment settings include the settings for the buyer agent, supplier agents and the broker agent.

4.1.1   Supplier Setting

The simulation contains six supplier agents and considers four attributes: price, electricity usage on weekdays, electricity usage on weekends and early withdrawal penalty. The detail contents of each supplier are presented in Table 1. Supplier agents use a bonus to attract buyer agents to purchase their electricity. In particular, five of the six supplier agents offer a bonus for the buyer agent and the satisfaction degrees of a bonus for 'gift' and 'free sign up fee' are set as 80 and 10 %, respectively.

4.1.2   Broker Setting

All supplier agents agree that if their electricity is bought by B_1 through BR_1, BR_1 will receive a reward value called r from supplier agents. In this experiment, the reward is calculated as

$$r = price \times 10\%$$

(4)

It can be seen that if there are more than one supplier agent, which satisfy B_1's requirements, BR_1 will choose a supplier with the largest reward to BR_1.

Table 1 Some electricity suppliers

Supplier	Price (AUD/KW)	Electricity usage on weekdays (KW)	Electricity usage on weekends (KW)	Early withdrawal penalty	Sale off
S1	1.40	270	360	No	No bonus
S2	0.70	200	290	Yes	Gift
S3	0.71	240	400	No	Free sign up fee
S4	0.80	245	320	No	Free sign up fee
S5	0.89	229	350	No	Gift
S6	0.98	248	420	No	Free sign up fee

4.1.3 Buyer Setting

The buyer agent B_1's requirements are presented as follows. B_1's concession threshold λ is set to a value (50 %) and four considered attributes in B_1's EER are price, electricity usage on weekdays, electricity usage on weekends and early withdrawal penalty. B_1's acceptability threshold is as 95 % and based on B_1's acceptability threshold, each constraint value of attributes is displayed in Tables 2, 3, 4 and 5, respectively. In addition, the priority degrees of price, electricity usage on weekdays, electricity usage on weekends and early withdrawal penalty are set to 3, 2, 1, and 4, respectively. Thus, B_1's EER is shortly presented as follows.

$$EER = \begin{pmatrix} & \text{Electricity usage} & \text{Electricity usage} & \text{Early withdrawal} \\ \text{Price} & \text{on weekdays} & \text{on weekends} & \text{penalty} \\ ---- & ----- & ----- & ----- \\ \text{under } 0.7 & \text{under } 200 & \text{under } 300 & no \\ 3 & 2 & 1 & 4 \end{pmatrix} \tag{5}$$

Table 2 Satisfaction degree of price

Price (AUD/KW)	Satisfaction degree (%)
Under 0.7	100
0.7–1.0	90
1.0–1.3	80
1.3–1.6	70
1.6–1.9	60
1.9–2.2	50
Above 2.2	40

Table 3 Satisfaction degree of electricity usage on weekdays

Electricity usage on weekdays (KW)	Satisfaction degree (%)
Under 200	100
200–220	90
220–240	85
240–260	80
260–280	70
280–300	60
Above 300	50

Table 4 Satisfaction degree of electricity usage on weekends

Electricity usage on weekdays (KW)	Satisfaction degree (%)
Under 300	100
300–400	70
Above 400	30

Table 5 Satisfaction degree of early withdrawal penalty

Early withdrawal penalty	Satisfaction degree (%)
No	100
Yes	0

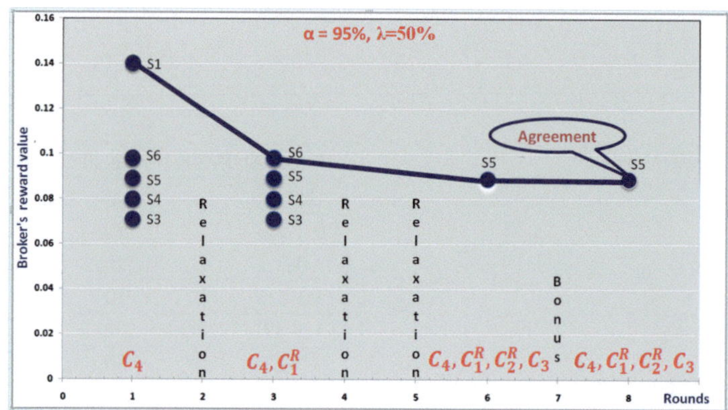

Fig. 2 The experimental result

4.2 Experiment Results

In this subsection, we illustrate the experimental results on the trading process between a buyer agent and supplier agents through a broker agent for supplier selection in the power market.

In the experiment, B_1's acceptability threshold is set at a high value (95 %). This means that it is difficult for the trading process between B_1 and $\mathbf{S}$ through BR_1 to achieve an agreement without any relaxation of requirements. Thus, our approach is useful for overcoming this difficulty. In particular, B_1 uses a relaxation when BR_1 cannot find any supplier agent, which satisfies B_1's requirements. Supplier agents offer a bonus program to attract B_1 to purchase their electricity and BR_1 selects the most suitable supplier for B_1.

The experimental result is illustrated in Fig. 2. From Fig. 2, we can see that the agreement was achieved through using 8 rounds. The relaxation was applied in rounds 2, 4, and 5 because BR_1 could not find any supplier agent, which could satisfy B_1's requirements. After the relaxation was used in round 5, BR_1 found that S_5 could meet B_1's requirements and required B_1 to verify whether S_5 was acceptable. Although S_5 satisfied all constraints of B_1, the agreement was not achieved because B_1's acceptability degree was 92.5 % for S_5 which was less than B_1's required acceptability threshold of 95 % in round 6. So, B_1 required BR_1 to find other supplier agent. Then, BR_1 found S_5 again with offered bonus and required B_1 to verify whether S_5 could be acceptable in round 7. B_1 calculated the acceptability degree for S_5 with the offered bonus. The acceptability degree of S_5 was acceptable and the agreement was achieved in round 8.

The explanation of such results is (1) The buyer agent used relaxation three times to achieve an agreement with the acceptability threshold $\alpha=95$ %. If the relaxation was not carried out, the trading process was terminated in round 2. (2) Supplier agents used a bonus policy to attract the buyer agent to purchase their electricity in round 7. Thus, the experimental result convinces us that the buyer's relaxation strategy, suppliers' bonus policy and the broker's supplier selection are able to successfully achieve a trading process.

4.3   Discussions

There has been a lot of previous work on regarding the indirect interaction between buyer agents and seller agents through broker agents in e-markets. Sarma et al. [11] analyzed market behavior in large networks where buyer agents do not know seller agents and vice-versa. All trading processes between seller agents and buyer agents depend on broker agents. Although they proposed polynomial time algorithms to compute equilibria in networks, their proposed algorithms are only suitable for simple situations in e-markets. The difference between Sarma's work and our work is that all trading processes in our approach are related to three parties involving buyer agents, supplier agents and broker agents. Supplier agents offer a bonus policy to attract buyer agents to purchase their electricity supply. In our approach, broker agents will achieve supplier agents' reward if broker agents sell supplier agents' electricity supply to buyer agents, while Sarma's work does not pay attention to a bonus policy from seller agents to buyer agents and broker agents. Blume et al. [2] modeled the trading phenomenon related to buyer agents, seller agents and trader agents. In their model, trader agents set price strategies, and then buyer agents and seller agents react to the proposed price. Although their model proposed price strategies, it did not incorporate in many features of real markets. The novelty of our approach is that broker agents act as middlemen to find a potential supplier agent to meet the buyer agents' requirements, while Blume's work does not pay attention to how to match buyer agents and seller agents through broker agents.

Conclusion and Future work

This paper proposed a relaxation strategy with fuzzy constraints to select a potential supplier agent for a buyer agent in the power market through a broker agent. The strategy contains three main components: supplier selection, a fuzzy constraint relaxation and decision making. The proposed approach is novel because (1) the trading process between the buyer agent and supplier agents through the broker agent is successfully solved based on fuzzy constraints for multiple attributes of the buyer agent's requirements and power supplier agents; and (2) the buyer agent's relaxation is applied to select a potential supplier agent through the broker agent when the buyer agent's requirements cannot be met by any supplier agents. Also, the experimental results demonstrate the good performance for supplier selection in the power market.

Further work is needed to test the proposed strategy in a real world application and to develop comprehensive strategies with fuzzy constraints to consider the relationships of the buyer agents and the broker agents.

References

1. Bichler, M.: The Future of e-markets: Multidimensional Market Mechanisms. Cambridge University Press, Cambridge (2001)
2. Blume, L.E., Easley, D., Kleinberg, J., Tardos, E.: Trading networks with price-setting agents. Games Econ. Behav. **67**(1), 36–50 (2009)
3. Fatima, S., Wooldridge, M., Jennings, N.R.: An analysis of feasible solutions for multi-issue negotiation involving nonlinear utility functions. In: Proceedings of The 8th International Conference on Autonomous Agents and Multiagent Systems, vol. 2, pp. 1041–1048. International Foundation for Autonomous Agents and Multiagent Systems (2009)
4. Gale, D.M., Kariv, S.: Financial networks. Am. Econ. Rev. **97**(2), 99–103 (2007)
5. Ketter, W., Collins, J., Reddy, P., Flath, C., Weerdt, M.: The 2012 power trading agent competition. ERIM Report Series Reference No. ERS-2012-010-LIS (2012)
6. Luo, X., Jennings, N.R., Shadbolt, N., Leung, H., Lee, J.H.: A fuzzy constraint based model for bilateral, multi-issue negotiations in semi-competitive environments. Artif. Intell. **148**(1), 53–102 (2003)
7. Nicolaisen, J., Petrov, V., Tesfatsion, L.: Market power and efficiency in a computational electricity market with discriminatory double-auction pricing. IEEE Trans. Evol. Comput. **5**(5), 504–523 (2001)
8. Ren, F., Zhang, M.: Desire-based negotiation in electronic marketplaces. In: Innovations in Agent-Based Complex Automated Negotiations, pp 27–47. Springer (2010)
9. Ren, F., Zhang, M., Fulcher, J.: Bilateral single-issue negotiation model considering nonlinear utility and time constraint. In: New Trends in Agent-Based Complex Automated Negotiations, pp. 21–37. Springer (2012)
10. Rubinstein, A., Wolinsky, A.: Middlemen. Q. J. Econ. **102**(3), 581–593 (1987)

11. Sarma, A.D., Chakrabarty, D., Gollapudi, S.: Public advertisement broker markets. In: Internet and Network Economics, pp. 558–563. Springer (2007)
12. Wijaya, T.K., Larson, K., Aberer, K.: Matching demand with supply in the smart grid using agent-based multiunit auction. In: Fifth International Conference on Communication Systems and Networks (COMSNETS), pp. 1–6. IEEE (2013)

Idea Discovery: A Context-Awareness Dynamic System Approach for Computational Creativity

Hao Wang, Yukio Ohsawa, Xiaohui Hu, and Fanjiang Xu

Abstract Chance Discovery is an emerging interdisciplinary area focusing on detecting rare but important events or situations through human-computer interaction for human decision making in complex real world. In this paper, we propose a context-awareness systematic approach named Idea Discovery for idea cultivation, construction, integration and evaluation, extending Chance Discovery into a dynamic discovery process. Idea Discovery process contains two key components: (1) mining algorithms for context construction in the computer process. (2) context-based creativity support techniques for increased awareness in the human process. A case study in an automobile industry for product innovation has validated the effectiveness of this approach.

Keywords Chance discovery • Context-based creativity support • Idea discovery

1 Introduction

Chance Discovery is a relatively new research field as an extension of text-based knowledge discovery, which is a human-computer interaction process to detect rare but important chances for decision making. A chance means to understand an unnoticed event or situation which might be uncertain but significant for a decision [7]. A core visualization tool called KeyGraph can generate context map to aid human's value cognition the double-helix process of chance discovery. In fact, KeyGraph is a keyword extraction algorithm from a single document using co-occurrence graph [6]. That is, a document is represented as a graph where each node corresponds to a term and each edge means the co-occurrence of two terms. Based on the segmentation of a graph into clusters, Keygraph extracts keywords by selecting the terms which strongly co-occurs with clusters. Figure 1 shows

H. Wang (✉) • X. Hu • F. Xu
Institute of Software, Chinese Academy of Sciences, Beijing, China
e-mail: wanghao@iscas.ac.cn

Y. Ohsawa
The University of Tokyo, Tokyo, Japan

© Springer Japan 2015
Q. Bai et al. (eds.), *Smart Modeling and Simulation for Complex Systems*, Studies
in Computational Intelligence 564, DOI 10.1007/978-4-431-55209-3_7

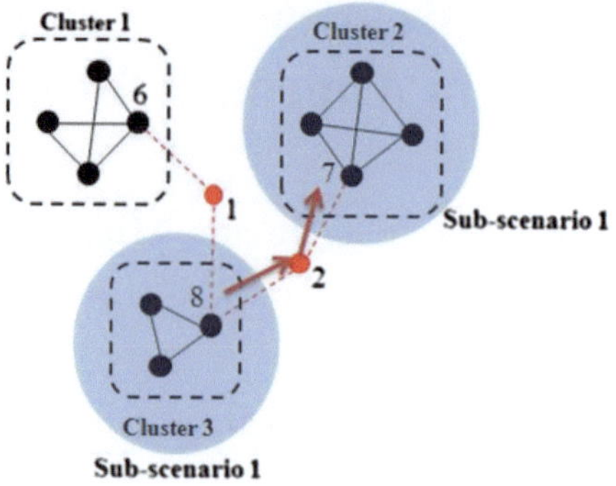

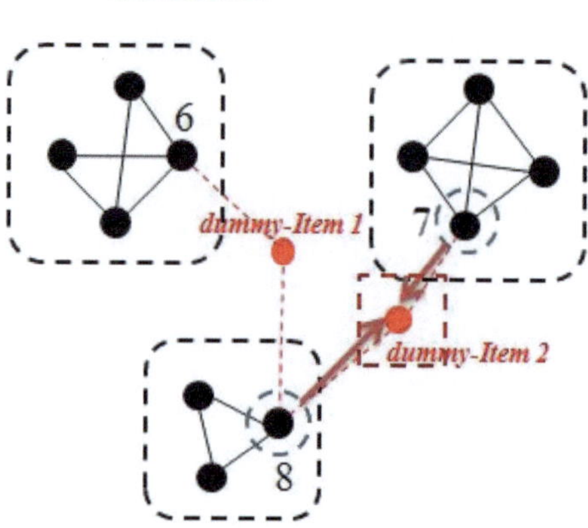

Fig. 1 A sample of KeyGraph context map

Fig. 2 Creating chances through combinational thinking

a context map visualized by KeyGaph. The red nodes are considered as chance candidates because they act as a bridge linking different sub-contexts. Human are required to understand and interpret the importance of chance candidates that can make the situation transfer from one sub-context to another. Later, Ohsawa [4] proposed a breaking-through method named data crystallization where dummy nodes representing invisible events are inserted into the processed text data, and then new data is visualized by KeyGraph. However, the complex algorithm and graph obtained were hard for users to understand, thus [3] subsequently present a new method, human-computer interactive annealing, for revealing latent structures and discovering dark events. Based on Chance Discovery, Hong [1] proposes an interactive human-computer process model called Qualitative Chance Discovery Model (QCDM) to extract more accurate data representation in context map for decision making on potential chances.

In recent years, a context map generated by KeyGraph with data crystallization has been applied as the game board in Innovators Market Game ®(IMG) and Innovators Marketplace ®(IM) for innovative chance discovery [5, 8, 9]. In particular, human cognition to KeyGaph context map has been expanded from sensing bridge-like chances to creating chances by combinational thinking, see Fig. 2. Wang and Ohsawa [13] have proposed a systematic approach with IMG and applied it in creative products designing. To further improve IMG, a customer-centric creativity support technique called 4W-IMG, has been proposed [11].

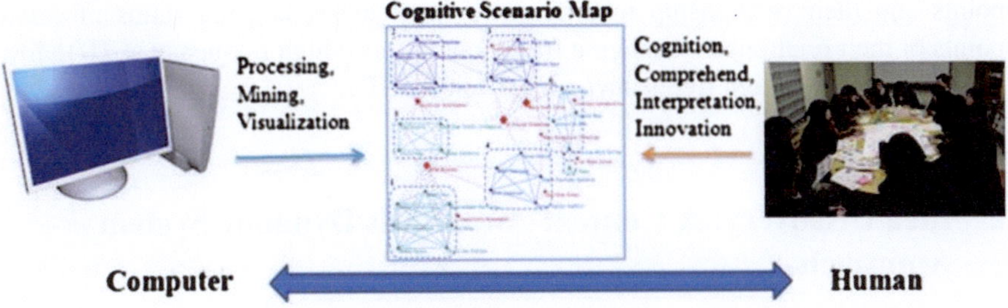

Fig. 3 A basic framework of discovery pattern

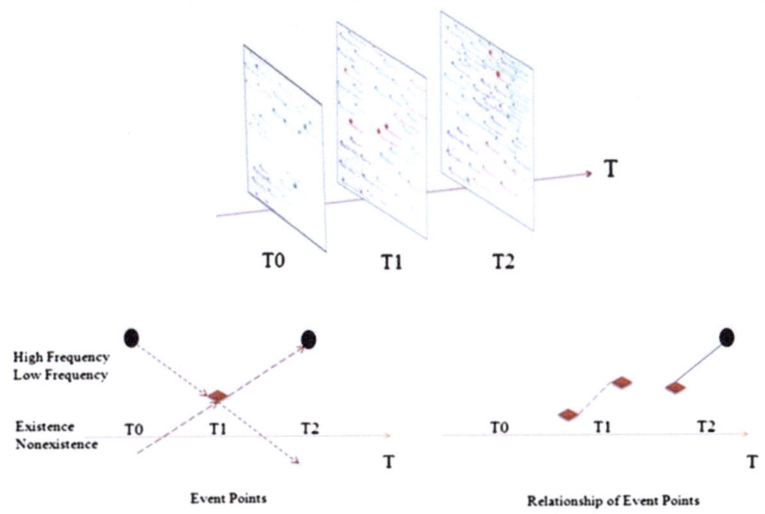

Fig. 4 The dynamic changes of event points and their relationship

Even so, the previous researches on text-based knowledge discovery or chance discovery both combine two processes: one is the computer process of context mining and visualization; the other is the human process of cognition, understanding, interpretation and innovation. Figure 3 shows a basic framework of current discovery pattern.

Dynamic discovery is commonly used by enterprises to evaluate and understand technology trends through patent data analysis, and eventually achieve a strategic advantage. Kim et al. [2] proposes a patent map visualization method, but it fails to automatically track the changes of patent trends in different periods. Shih et al. [10] overcome the problem and propose a patent trend mining method to automatically identify changes of patent trends without specialist knowledge.

In this paper, dynamic discovery focuses on discovering the changing trends of event points and their relationship/links through the comparison of two consecutive context maps in time series. Figure 4 shows the dynamic change process of event points and their relationship in different context maps, i.e., the status of event points or their relationship changing from nonexistence/high frequency at T0 to low frequency at T1 to high frequency/ nonexistence at T2.

2 Idea Discovery: A Context-Awareness Dynamic System Approach

Idea Discovery, as an extension and evolution of chance discovery, is a dynamic and sustainable process for high-quality ideas cultivation, construction, generation and evaluation through human-computer-human interaction. Idea Discovery not only works on rare and important event points, but also focuses on latent and significant event relationship and the dynamic changes of these events and their relationship. Therefore, Idea Discovery is committed to digging up latent information (event points and their relationship) and its dynamic changes through static and dynamic discovery, for more actionable ideas creation, integration and evaluation. Figure 5 reveals a dynamic model of Idea Discovery process and the details are presented as below:

Step 1: Data gathering and preprocessing. Determine the objective of task and select relevant data. And then text data is preprocessed into a group of basket data sets denoted by D, each row of which is a basket data set. For example, $P1, P2, P3$ and $P4$, these four items constitute a basket data set.

$$D = P1, P2, P3, P4$$

$$P2, P7, P5$$

$$P3, P10, P6, P9, P5$$

$$\ldots\ldots$$

Step 2: Context map construction. Preprocessed data D is mined by a mining algorithm and the result is finally visualized into a context map. Here, we

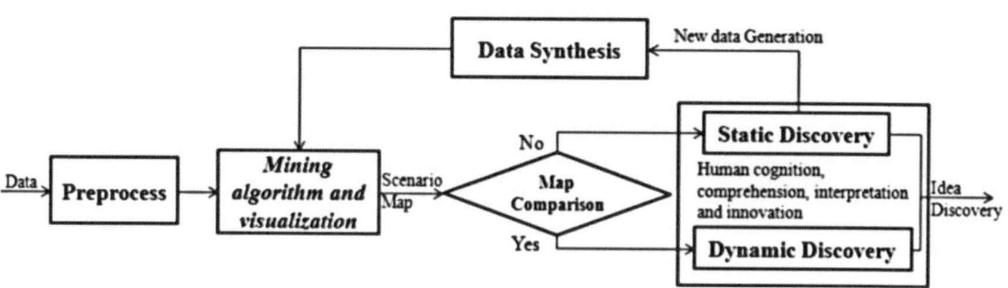

Fig. 5 Idea discovery dynamic discovery process

Table 1 Static and dynamic discovery in Idea Discovery

	Static discovery	Dynamic discovery
Context Map	Single	Multiple
Thinking Mode	Combination/Analogy/Reasoning	Dynamic Observation
Observed Objects	Event points and relationship (Nodes and links in map)	Changes of event points and relationship (Status changes of nodes and links in map)

apply *IdeaGraph* algorithm. *IdeaGraph* is a novel algorithm to generate a rich context map for human cognition, comprehension, interpretation and innovation. *IdeaGraph* not only works on discovering more rare and significant events, but also focuses on uncovering latent relationship among events. The details of *IdeaGraph* algorithm will be presented in the next section.

Step 3: Context maps comparison. This step refers to a choice of discovery pattern, static discovery and dynamic discovery. If a dynamic discovery is needed, current and previous context map will be compared for observing dynamic changes of latent information. As mentioned previously, the dynamic changes refer to an event point or an event relationship changing from nonexistence to low frequency to high frequency, and vise verse. If a static discovery process is chosen, the context map will be provided for static discovery.

Step 4: Static and dynamic discovering. Table 1 illustrates there are two kinds of discovery pattern in Idea Discovery.

One pattern is the static discovery that user groups obtain value cognition from context map and come up with new ideas by combinational, analogical and reasoning thinking. Reasoning thinking needs directed context map, so *IdeaGraph* algorithm is also used to generate directed map as well as undirected one. The other is dynamic discovery that human conceives new ideas through dynamic observation. Here, a tool is needed to track dynamic changes of each event point and their relationship by comparing before and after context maps.

In the static discovery, a group brainstorming or innovation game can be used to accelerate the discovery process. Here, a Web-based creativity support system called *iChance* is applied for collective intelligence [11, 12]. A group of new basket data sets D is output from *iChance*. Each basket set in D indicates how a new idea is created. For instance, a user creates a new *Idea*1 with consideration of event $P2$, event $P5$ and *Cognition*1 that is proposed by other users. In fact, an idea is a meta-cognition through value synthesis.

$$D' = P2, P5, Cognition1, Idea1$$

$$P2, P3, P4, Idea2$$

$$P10, P6, P9, Cognition1, Idea3$$

$$\cdots\cdots$$

Step 5: Data Synthesis. The synthesized data Syn_D is obtained by adding D to the end of D.

$$SynD' = P1, P2, P3, P4$$
$$P2, P7, P5$$
$$P3, P10, P6, P9, P5$$
$$\cdots\cdots$$
$$P2, P5, Cognition1, Idea1$$
$$P2, P3, P4, Idea2$$
$$P10, P6, P9, Cognition1, Idea3$$

Step 6: Idea discovery process iteration. Return to Step 2 for a new round of idea discovery. *IdeaGraph* is applied again to generate a new context map, and dynamic discovery may be carried out by comparing current new context map with previous map.

3 IdeaGraph: An Algorithm of Exploiting More Latent Information for Context Construction

In this section, we introduce a human-oriented context construction algorithm called *IdeaGraph* for human's comprehension, interpretation and innovation. *IdeaGraph* not only works on discovering more rare and significant chance events, but also focuses on uncovering latent relationship among them [14]. Suppose that collected data has been preprocessed into ID.

$$ID' = item1, item2, item3, item4$$
$$item2, item7, item5$$
$$item3, item6, item9, item10, item5$$
$$\cdots\cdots$$

Figure 6 shows a context map construction process by *IdeaGraph* and the details are presented as below:

Step 1: Generating general clusters. The relationship between two items is measured by their conditional probability. That is, the relationship of any two items, I_i and I_j, is calculated by Eq. (1). And then the pairs whose $R(x, y)$ are greater than preset threshold r are linked by solid line in the graph G. In this way, general clusters are obtained and denoted by C_i. The high-frequency items which are greater than preset threshold f is represented by black nodes, while low-frequency ones are denoted by red nodes.

$$R(I_i, I_j) = P(I_i|I_j) + P(I_j|I_i) \tag{1}$$

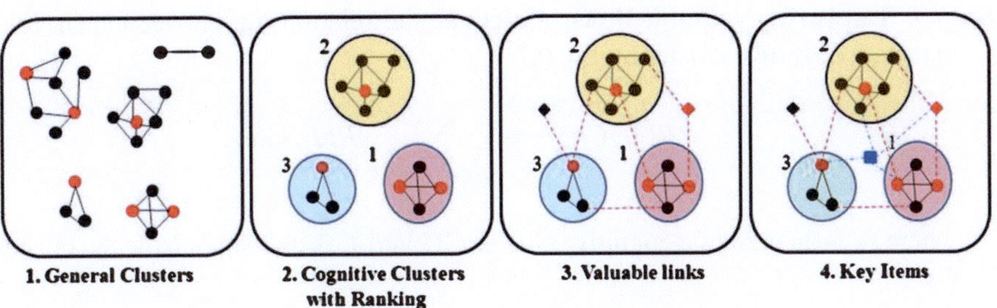

Fig. 6 The context map construction process using *IdeaGraph* algorithm

Step 2: Obtaining cognitive clusters. *Cognitive Cluster* is defined as a cluster contains rich information but should be small enough for human cognition and interpretation. To obtain cognitive cluster, *information* and *information density*, these two indicators are employed to quantify general clusters generated in Step 1.

The definition of *information* is the sum of $R(I_i, I_j)$ of all the edges in a general cluster. The *information density* is defined that the *information* of a cluster is divided by the number of items in the cluster. That means the *information density* of a cluster is the *information* of each item in this cluster. Thus the equations of *information* and *information density* are

$$Info(C) = \sum_{I_i, I_j \in C} R(I_i, I_j) \tag{2}$$

$$InfoDen(C) = \frac{Info(C)}{Ne} \tag{3}$$

where I_i or I_j is an item of a cluster C and N_e indicates the number of items in the cluster C.

Each general cluster is measured by the harmonic average of these two indicators. Equation (4) is derived from merging Eqs. (2) and (3). Therefore, each general cluster is measured by the harmonic average of these two indicators.

$$ClusterVal(C) = \frac{2Info(C)}{Ne + 1} \tag{4}$$

Equation (4) indicates it favors the cluster which has fewer items when two general clusters have the same *information*.

Therefore, all general clusters are ranked by their $ClusterVal(C)$ in a descending order and parts of them are chosen as cognitive clusters denoted by CC through picking up the top N_c clusters.

Step 3: Capturing valuable links. Calculate the relationship between each item and each cognitive cluster by Eq. (5).

$$PR(I_i, CC) = \sum_{I_i \notin CC, c_k \in CC} R(I_i, c_k) \tag{5}$$

where c_k is an item of a cognitive cluster CC and I_i is an item outside the cluster CC.

Then, item-cluster pairs are sorted and the top M_1 pairs are selected to be linked by red dot line. The new items are added if they dont appear in the graph G.

Step 4: Extracting key items. A key item is the item which has strong relationship with all the other cognitive clusters and newly added items in Step 3. The key value $Key(I)$ of an item is calculated by Eq. (6).

$$Key(I) = \sum_{i=0}^{N_c} PR(I, CC_i) + \sum_{I_k \notin CC, I_k \in G} R(I, I_k) \tag{6}$$

All items are sorted by their $Key(I)$ and the top N_k items are taken as key items which are added in the graph G if they dont exist.

4 A Case Study in Automotive Industry

We have successfully carried out a project in a famous auto company. The objective of the project is to explore Chinese users preference on human-machine interface (HMI) system for further development. HMI system may aid users simple access to all infotainment components, such as navigation, telephone, video/audio, entertainment information, etc. To improve user experience while driving, the company is eager to discover more potential business opportunities which help to create new ideas and strategies on new products or services.

We provide Idea Discovery systematic approach to achieve the goal of the company. Our main tasks are as below:

1. Analyzing current HMI situation in automotive industry. *IdeaGraph* is employed to mine and visualize the data from market investigation.
2. Identifying Chinese users preference on car HMI system. *iChance*, a creativity support system is applied by the company to explore and understand potential demands of customers.
3. Achieving more actionable ideas and strategies for further product development. The static discovery and dynamic discovery of Idea Discovery will help the company develop more actionable functions and strategies.

4.1 Implementation Procedure

An intelligence system named *Galaxy* has been developed to support the process of idea discovery [15]. *IdeaGraph* is implemented and integrated in *Galaxy* system. The procedure of project implementation is presented as follows:

Step 1: According to the project objective, we select relevant data - 96 question-naires from market investigation. The data is preprocessed into 650 basket data sets where each item represents a specific function of HMI. We also obtain valuable subjective data through a group brainstorming and eventually collect 733 basket sets and visualize them into an *IdeaGraph* context map, shown in Fig. 7.

Step 2: *IdeaGraph* context map is employed by *iChance* system for the company to make collaborative innovation with their customers on the Web [11, 12]. Five customers are invited to join in *iChance*, and one *Facilitator*, two *Experts*, two *Designers* from the company participate in *iChance* as well.

Step 3: We take the new data obtained from Step 2 to make data synthesis using *Galaxy*, generating a new data source with 1091 basket data sets. We visualize this new data source into a new context map shown in Fig. 8 and compare current map with previous one. Five group members of the company are required to observe dynamic changes of nodes and links and propose new ideas and strategies.

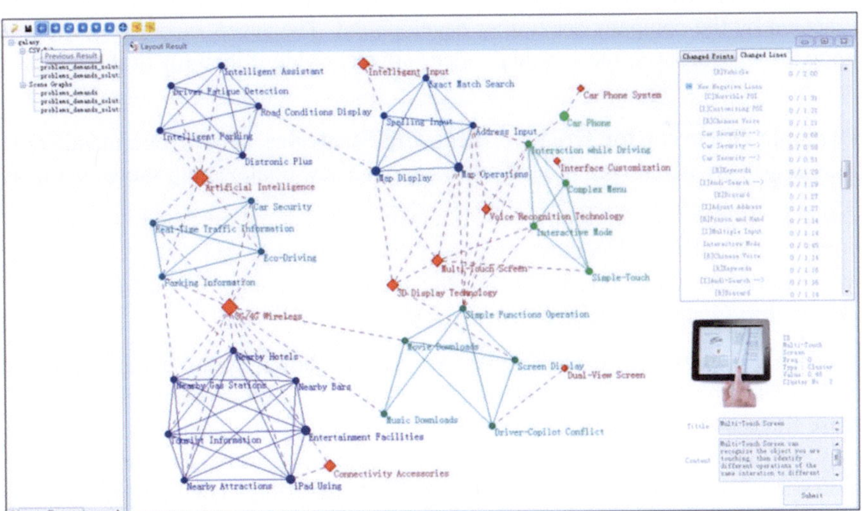

Fig. 7 *IdeaGraph* context map visualized by *Galaxy* with 733 basket data sets

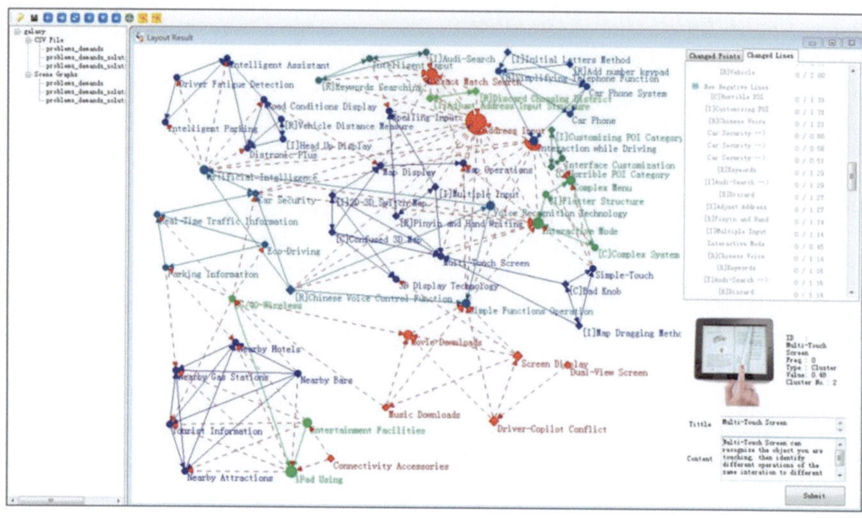

Fig. 8 *IdeaGraph* context map after data synthesis

4.2 The Results of Static and Dynamic Discovery

Static Discovery for Creative Ideas Creation. In the end, the company obtained ten product ideas from *IdeaGraph* context map. Nine of ten product ideas are accepted by the company for further development. The acceptance rate of ideas is up to 90 %. Moreover, these product ideas are clearly described for further product design.

Dynamic Discovery for Creative Ideas and Strategies. Five group members in the company firstly are required to understand the new context map shown in Fig. 8.

Dynamic Changes of Sub-contexts. As Fig. 9 shows, comparing with the previous context map, the new context map has three additional sub-contexts regarded as potential business opportunities in the future: (N1) *Simplified Structure*, (N2) *Personalization* and (N3) *Intelligence Search*.

Dynamic Changes of Nodes/Events. As Fig. 10 shows, group members found *Screen Display*, *Music downloads* and *The Conflict of Driver and Font Passenger*, these three nodes changing from high frequency to low frequency, which reveals they will be well solved by the ideas from static discovery. And another eight nodes that have changed from low frequency to high frequency indicate that they are probably emerging as potential business opportunities in the future. That is, *3D Display*, *Multi-Touch Screen*, *Interface Customerization*, *Car Phone System* will be popular in car owners. Whats more, group members found that they need change their design strategies to respond ever-changing customer demands and market opportunities. According to the changing nodes, such as *Artificial Intelligence, Voice*

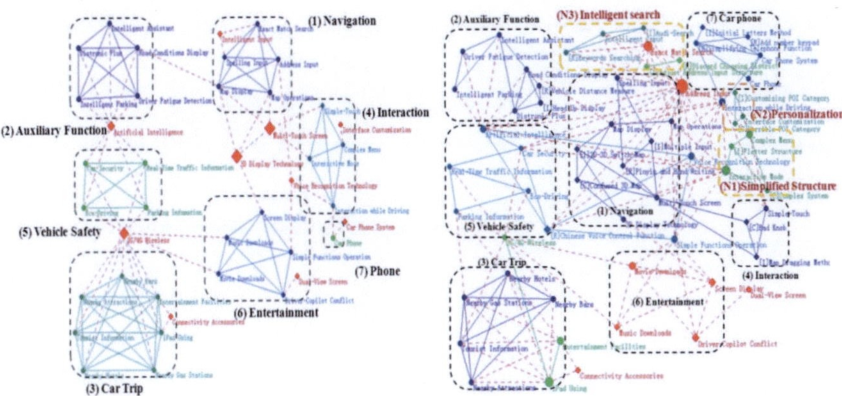

Fig. 9 The dynamic changes of sub-contexts

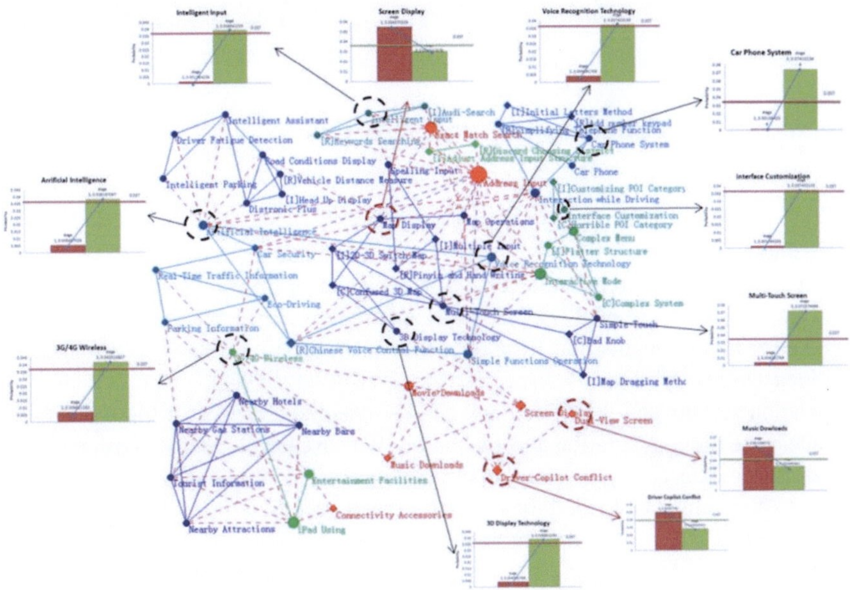

Fig. 10 The dynamic changes of links/relations between event nodes

Recognition Technology, *Intelligence Input* and *3G/4G Wireless*, group members suddenly realized a new strategy that *intelligence* and *interconnection* will be the trend of HMI product design and development.

Dynamic Changes of Links/Relations. In Fig. 11, we found that four nodes having more links with the other nodes are chosen to observe dynamic changes of links/relations: *Exact Match Search*, *Address Input*, *Interactive Mode* and *3G/4G Wireless*. In the end, another four ideas are further proposed.

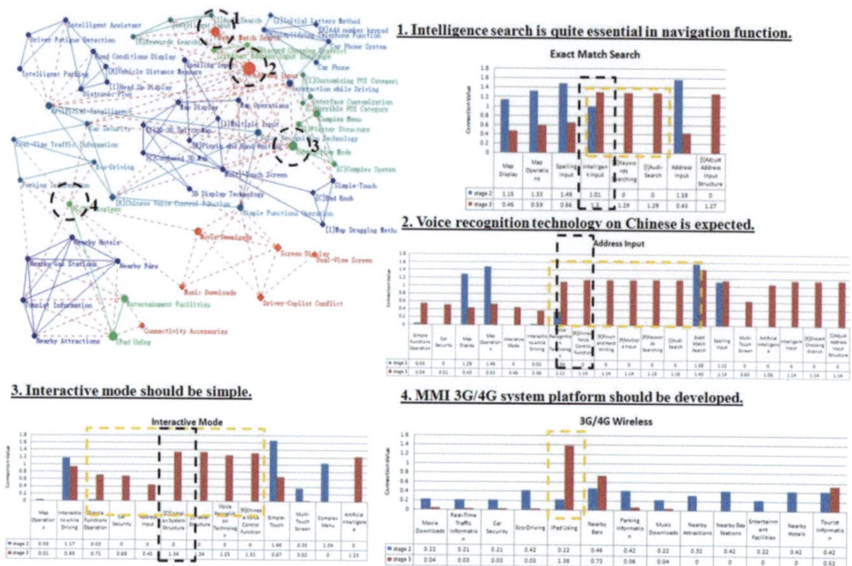

Fig. 11 Dynamic changes of links/relations between event nodes

Conclusion

In this paper, we proposed a context-awareness system approach called Idea Discovery for sustainable creativity support through human-computer and human-human interaction. Unlike traditional text-based discovery methods, Idea Discovery mainly depends on uncovering latent and important information (events and their relations) and its dynamic changes for high-quality ideas creation, integration and evaluation. We have developed an intelligence system called Galaxy to support idea discovery process. A case study in an auto company has verified the effectiveness of proposed method and system. We have helped the company achieve ten creative ideas, nine of which are accepted by the company for further development. Moreover, the dynamic discovery made the enterprise discover additional creative ideas regarded as potential business opportunities and timely develop their new strategies to respond ever-changing customer demands and market opportunities.

Acknowledgements This work is supported by Natural Science Foundation of China (61303164), National Basic Research Program of China (2013CB329305) and Beijing Natural Science Foundation (9144037). We would like to thank all participants for their great contributions to our case studies.

References

1. Hong, C.F.: Qualitative chance discovery - extracting competitive advantages. Inf. Sci. **179**(11), 1570–1583 (2009)
2. Kim, Y.G., Suh, J.H., Park, S.C.: Visualization of patent analysis for emerging technology. Expert Syst. Appl. **34**(3), 1804–1812 (2008)
3. Maeno, Y., Ohsawa, Y. Human-computer interactive annealing for dis-covering invisible dark events. IEEE Trans. Ind. Electron. **54**(2), 1184–1192 (2007)
4. Ohsawa, Y.: Data crystallization: chance discovery extended for dealing with unobservable events. New Math. Nat. Comput. **1**(3), 373 (2005)
5. Ohsawa, Y.: Innovation game as a tool for chance discovery. In: Proc. of twelfth international conference on Rough Sets, Fuzzy Sets, Data Mining and Granular Computing (RSFDGrC 2009) (pp. 59–66) (2009)
6. Ohsawa, Y., Benson, N. E., Yachida, M.: KeyGraph: Automatic indexing by co-occurrence graph based on building construction metaphor. In Proc. of the Advanced Digital Library Conference (pp. 12–18) (1998)
7. Ohsawa, Y., McBurney, P. (eds.): Chance Discovery. Springer, Berlin/Heidelberg (2003)
8. Ohsawa, Y.. Nishihara, Y.: Innovators Marketplace: Using Games to Active and Train Innovators. Springer, Berlin/Heidelberg (2012)
9. Ohsawa, Y., Okamoto, K., Takahashi, Y., Nishihara, Y.: Innovators Market Place as Game on the Table versus Board on the Web. In: Proc. of IEEE International Conference on Data Mining Workshops (ICDMW) (pp. 816–821) (2010)
10. Shih, W., Liu, D.R., Hsu M.L.: Discovering competitive intelligence by mining changes in patent trends. Exp. Syst. Appl. **37**(4), 2882–2890 (2010)
11. Wang, H., Ohsawa, Y.: iChance: a web-based innovation support system for business intelligence. Int. J. Organ. Collective Intell. **2**(4), 48–61 (2011)
12. Wang, H., Ohsawa, Y. (eds.): Data-driven innovation technologies for smarter business from innovators market game to iChance creativity support system. Advanced in Chance Discovery. Springer, Berlin/Heidelberg (2012)
13. Wang, H., Ohsawa, Y.: Innovation support system for creative product design based on chance discovery. Exp. Syst. Appl. **39**(5), 4890–4897 (2012)
14. Wang, H., Ohsawa, Y.: IdeaGraph: turning data into human insights for collective intelligence. In: Proc. of the 7th International Conference on Intelligent Systems and Knowledge engineering (ISKE 2012), Beijing (2012)
15. Wang, H., Ohsawa, Y.: Idea discovery: a scenario-based systematic approach for decision making in market innovation. Expert Syst. Appl. **40**(2), 429–438 (2013)

Hierarchical Scoring Rule Based Smart Dynamic Electricity Pricing Scheme

Shantanu Chakraborty and Takayuki Ito

Abstract Defining appropriate pricing strategy for smart environment is important and complex at the same time. In our work, we devise an incentive based smart dynamic pricing scheme for consumers facilitating a multi-layered *scoring rule*. This mechanism is applied between consumer agents (CA) to electricity provider agent (EP) and EP to Generation Company (GENCO). Based on the Continuous Ranked Probability Score (CRPS), a hierarchical scoring system is formed among these entities, CA-EP-GENCO. As CA receives the dynamic day-ahead pricing signal from EP, it will schedule the household appliances to lower price period and report the prediction in a form of a probability distribution function to EP. EP, in similar way reports the aggregated demand prediction to GENCO. Finally, GENCO computes the base discount after running a cost-optimization problem. GENCO will reward EP with a fraction of discount based on their prediction accuracy. EP will do the same to CA based on how truthful they were reporting their intentions on device scheduling. The method is tested on real data provided by Ontario Power Company and we show that this scheme is capable to reduce energy consumption and consumers' payment.

Keywords Demand response • Energy management • Incentive design • Scoring rule • Smart grid

1 Introduction

With the growing needs of environmental sustainability and continuing changes in electric power deregulation, smart grid becomes an inevitable choice for the society. While such grid infrastructure in mind, houses started to adopt devices which can be controlled, maintained, monitored and even scheduled as necessity calls. Smart house technology used to make all electronic devices around a house act "smart" or more autonomous. Most of the important appliances in the future will take advantage of this technology through home networks and the Internet. Such feature

S. Chakraborty (✉) • T. Ito
Nagoya Institute of Technology, Nagoya, Japan
e-mail: s-chakraborty@cp.jp.nec.com; shantanu.chakraborty@nitech.ac.jp

© Springer Japan 2015

Q. Bai et al. (eds.), *Smart Modeling and Simulation for Complex Systems*, Studies in Computational Intelligence 564, DOI 10.1007/978-4-431-55209-3_8

of smart grid is a way for ordinary electronics and appliances to communicate with each other, consumers and even energy provider (EP). Recently, smart pricing has attracted much attention as one of the most important demand-side management (DSM) strategies to encourage users towards consume electricity more wisely and efficiently [12].

On different note, *scoring rule* was defined [4] in order to numerically measure up the actual realization of a probabilistic event which was forecasted ahead. Moreover, it binds the assessor to make a careful prediction and hence truthfully elicit his/her private preferences. Which is why, *scoring rule* has been applied successfully while designing truthful incentive mechanism in diverse applications such as voting rules [13]. Strictly proper scoring rules can be employed by a mechanism designer to ascertain that agents accurately declare their privately calculated distributions, reflecting their confidence in their own forecast. The applicability of scoring rule is being investigated in field of smart-grid. For instance, [5] presented a methodology for predicting aggregated demand in smart-grid. Now, using smart devices, consumers (actually a consumer agent, refereed as CA hereafter, will be responsible to take such decision in conjunction with smart-meter) can respond to day-ahead dynamic pricing signal by effectively and intelligently managing and scheduling devices, thereby flattening out peak demand and achieving better resource utilization. This paper presents a hierarchical scoring rule based payment mechanism for CA provided by the EP and GENCO in response to the dynamic day-ahead time dependent pricing. The consumers will be rewarded a discount on the price to measure up how well they predict the shifting the devices/loads towards the lower demand (lower price as well) periods. These rewards are again a fraction of the discount which were provided by GENCO to the corresponding EP depending on EP's prediction of required energy demand. The reward mechanism is based on a strictly proper scoring rule. The scoring rule is chosen to reflect to work with continuous variable (the normal distribution, as in the proposed method) and measure up how accurate the prediction could be. The Continuous Ranked Probability Score [7] possess such characteristics.

2 Incentive Based Dynamic Pricing: System Model

GENCOs and EPs are responsible to determine electricity pricing. GENCOs make revenue by selling energy to the distributers (in our context, EP) based on their demand while distributers provide that energy to consumers.[1] A supply-demand chain is thus formed among these entities. For that model in hand however, it is critically important to have a sophisticated smart pricing scheme that will take advantage of the DSM technique as well as incentivize the CAs to schedule smart devices in order to reduce the total demand.

---

[1]Since, the scenario takes place in smart-grid infrastructure, we assume that all the consumers participating are equipped with smart devices.

As a mechanism deign to incentivize agents (both the CAs and EPs) for providing private probabilistic information accurately (truthfully) and to the best of their forecasting ability, scoring rule is being applied in this model. Interestingly, this scenario in particular coincides with DSM strategy where consumer responses to demand by shifting their device to lower price periods. Therefore, EP incentivizes consumers not only based on their prediction accuracy but also on the question of whether they shifted such loads to lower priced periods. Strictly proper scoring rules can be employed by a mechanism designer to ascertain that agents accurately declare their privately calculated distributions, reflecting their confidence in their own forecast. The details flow of information and task assignments are pointed in Fig. 1. As we can see, GENCO will send the price informationas a signal to EP.

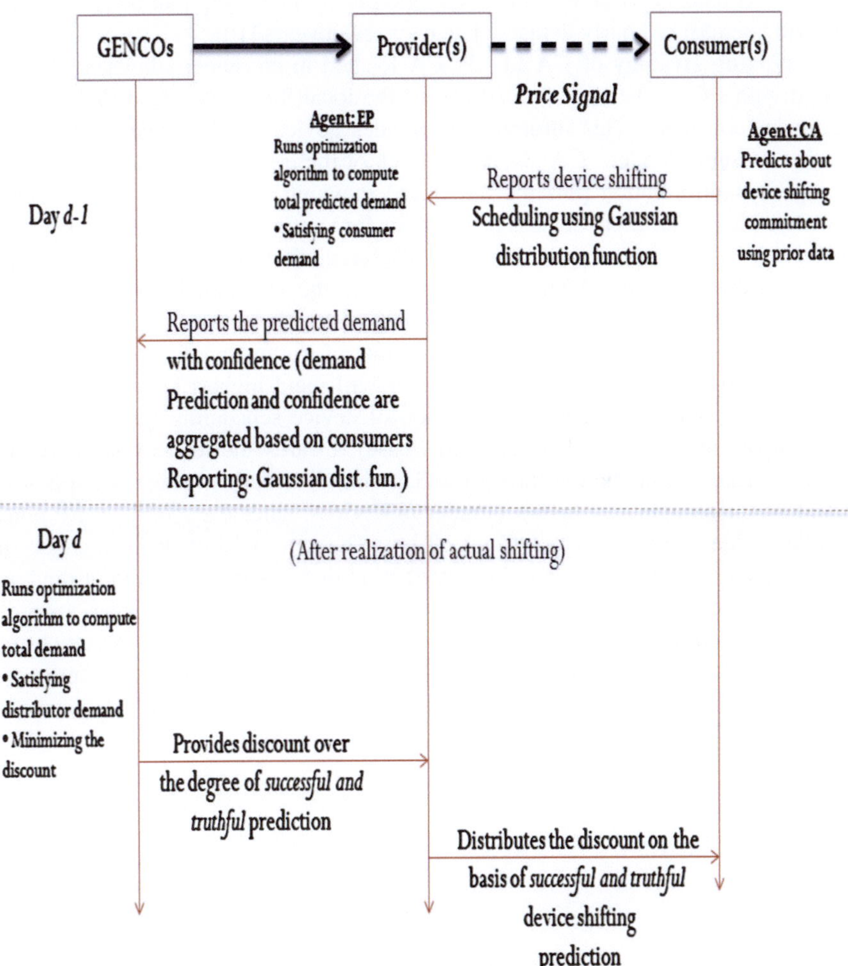

Fig. 1 Information flow of the GPC model

The price signal is typically determined based on the generation costs of electricity.[2] Although this model does not include the price determination mechanism, we assume that in dynamic pricing environment, the signal follows the demand. Which is, the price is higher when the demand is higher and its lower when demand is lower. The price signals are then conveyed to CAs via EPs. One thing can be noted that, one EP can provide energy to one or more consumers while one GENCO can also serve one or more EPs. Since, this model assumes a dynamic 'day-head' pricing signal, CAs receive their prices one day in advance. So, CAs can schedule their device usages for the upcoming day into the lower price periods. Lets say, the demand in each period i is D_i. The demand D_i in each period is assumed to be roughly the same each day due to repeated daily patterns in electricity demands (e.g. period 1 has the same demand on Monday, Tuesday, etc.). So, the aggregate demand over each day is usually constant. This assumption is verified using real traces from an Ontario operator of hourly demand data over seven years [10].

Reporting strategy of CA to EP: CA located in consumer's household integrated with ECC. Therefore, it can access the local information and data of that particular consumer. This information includes device usage schedule, duration, energy consumption, etc. CA also keeps track of the previous schedule prediction. Using such information plus the day-ahead dynamic pricing, CA makes a pre-schedule plan of different devices based on its forecasting accuracy and consumer's preferences. However, it will report EP the prediction confidence in a form of Gaussian distribution and tentative schedule of the assigned devices. For each device, CA calculates its uncertainty over the error it expects to make using a statistical model of random errors. An example of truthful prediction is shown in Fig. 2. As shown in the figure, a consumer will yield highest benefit where the in-house CA makes an accurate prediction of device scheduling for two devices (microwave and vacuum cleaner; in this case) towards the lower priced period. So, CA makes its prediction through a Gaussian distribution. This assumption is based on the sampling of higher number of devices, since eventually an EP must handle a wide range of devices. Central Limit Theorem (CLT) tells us in a case of wide range of events, the probability distribution that describes the sum of random variables tends towards a Gaussian Distribution as the sum approaches towards infinity. According to that, we can say that, in the perspective of EP, reporting error in the form of Gaussian Distribution is valid and portrays some form of accuracy.

---

[2]In our model, we assume that GENCOs operate on multiple plants of different types, such as coal, hydro and nuclear. Therefore, pricing signal could be a function of statistical forecast of historical price and the payment of EPs to purchase energy from generation companies.

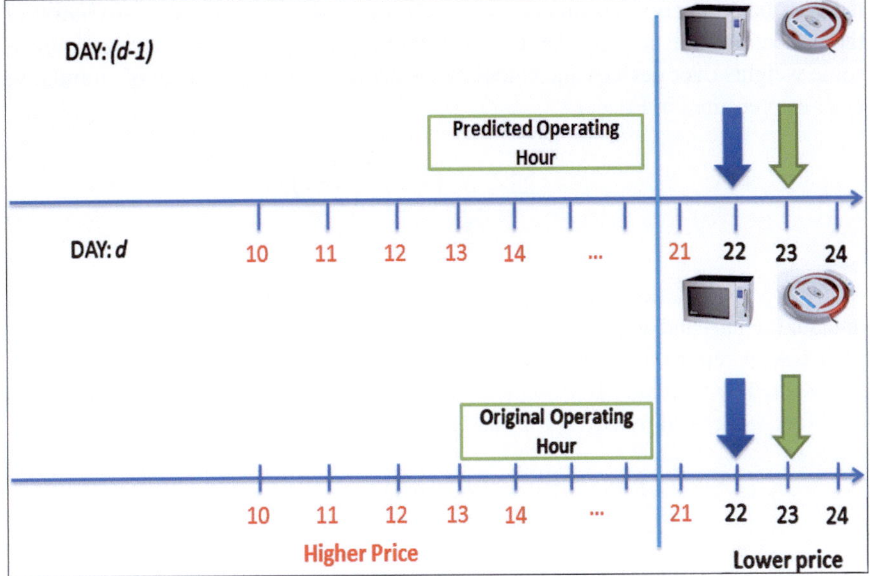

Fig. 2 Truthful prediction in device scheduling

2.1 Continuous Ranked Probability Score (CRPS)

"Scoring rules" have been used to elicit true information from a forecaster. It can be effectively utilized in an agent-driven scenario such as the model proposed in the paper. Given the day-ahead price signal, the CA makes a prediction for the next day. At the same time, however, if it sticks to the prediction of shifting devices to the lower priced periods, the utility provider (EP) can also make wise prediction about the required energy. As the figure suggests, the proposed model is dealing with devising scoring rule for multi-item prediction.

Therefore, conventional scoring rule with single item, such as *Brier score* [1] is of no use for this mode. In order to rightfully incentivize the consumer agents to make their prediction of device shifting for multiple items rightfully; the continuous ranked probability score (CRPS) is applied [4]. CRPS is a *strictly proper*[3] scoring rule that also ensures the truthfulness of the reporting [4] as well as works with continuous variable In the proposed method, Gaussian Distribution is used to model the consumers device shifting prediction and associated confidence. The usage of CRPS is investigated before in distributed power system operation to rightfully score

---

[3] A strictly proper scoring rule is a proper scoring rule where the forecaster has a unique maximum of its expected utility while a proper scoring rule has more than one maximal.

the distributed energy resource(s) [11]. CRPS is able to measure the *closeness*[4] of the prediction. Since, every device has different priority level of usage, we impose some weights over devices and calculate the actual weighted average of cumulative error as presented in Eq. (1)

$$\delta_u = \frac{\sum_{d \in DV_u} \left(W_d \left| \frac{P_d^a - P_d^p}{P_d^p} \right| \right)}{\sum_{d \in DV_u} P_d^p} \tag{1}$$

where P_d^a and P_d^p describe the prices of energy when the devices are operated at hours a (actual) and p (predicted). DV_u is the set of devices for CA, u. Lets assume, each CA, u reports its relative prediction error in a form of uncertainty over it, represented by *Gaussian Distribution Function* $\mathcal{N}(\mu = 0, \sigma_u^2)$. The reward score is therefore, generated by CRPS for that particular u is defined as Eq. (2)

$$\begin{aligned} CRPS(\mathcal{N}(\mu = 0, \sigma_u^2), \delta_u) \\ = \sigma_u \left[\frac{1}{\sqrt{\pi}} - 2\varphi \left(\frac{\delta_u}{\sigma_u} \right) - \frac{\delta_u}{\sigma_u} \left(2\Phi \left(\frac{\delta_u}{\sigma_u} \right) - 1 \right) \right] \end{aligned} \tag{2}$$

where the probability density function and cumulative distribution function for *Gaussian Distribution Function* are denoted as φ and Φ, respectively. The notation $CRPS(\mathcal{N}(\mu = 0, \sigma_u^2), \delta_u)$ can be simplified using $CRPS(\sigma_u^2, \delta_u)$. Accurate prediction of device schedule does not necessarily incentivize an agent (CA or EP) to truthfully report its intentions regarding device shifting. For instance, a CA can misreport about shifting period of a particular device (or group of devices) to a higher priced period while in actual it shifts the device in a lower priced time. Therefore, although the CA will lose some discounts by incorrect prediction, it will gain benefit by shifting device(s) in lower priced period. Moreover, since CRPS is also capable of measure how close the prediction was, CA always gets some amount of discount unless the prediction is highly erroneous. In order to incorporate such scenario and strictly incentive the CA for reporting its true prediction, the scoring rule needs to be revised. The scoring rule for a particular CA, u is redefined as,

$$SR_u = CRPS(\sigma_u^2, \delta_u) \times \left[\min_{d \in DV_u} \frac{4 + (P_d^a - P_d^p)}{e^{4(P_d^a - P_d^p)}} \right]^+ \tag{3}$$

where $[x]^+$ signifies *maximum*$(0, x)$. This above function ensures that any misreporting by CA will generate a significantly low score. Moreover, whenever a CA agent is trying to be extra-clever by achieve both lower price and discount while misreporting the intensions, the above equation penalize them by providing no discount.

---

[4]*Closeness* determines how close a prediction to realized event where higher value is assigned to close prediction.

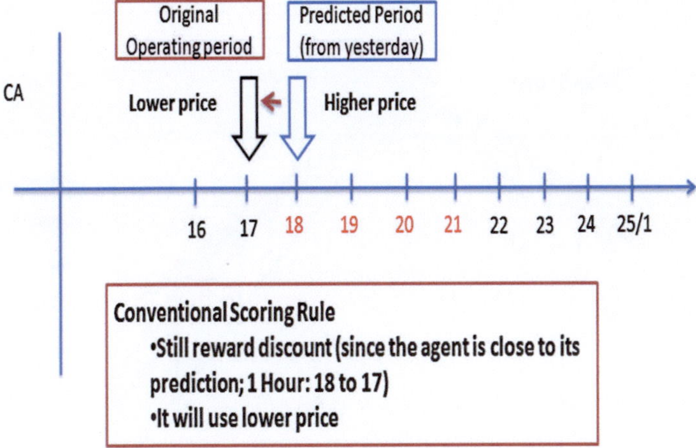

Fig. 3 Fault in conventional scoring rule

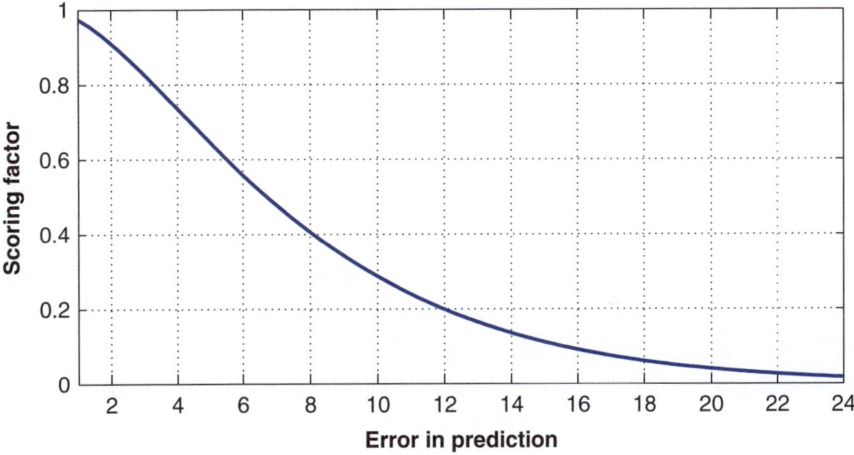

Fig. 4 Scoring pattern for scoring factor used in Eq. (3)

For example, consider a case (demonstrated in Fig. 3 where a CA wants to use a device at period 17 (lower price) and misreports that it will use it at period 18 (which is higher priced period than period 17, i.e. $(P_d^a - P_d^p) < 0$). Therefore, although it gets a less score for misreporting, it would appear that it will be compensated by lower price in period 17. Placing this situation in Eq. (3), it will generate a score of zero since for that device, the later part of Eq. (3) will produce a zero value. Hence, any CA who misreports of its true intention about shifting any device, will get a very significantly low score. Moreover, it also forces the agents to stay as close as possible to their predicted schedule, which ensures the *closeness* property. The 2nd part of Eq. (3) is plotted in Fig. 4 to show the effect of relative factor against

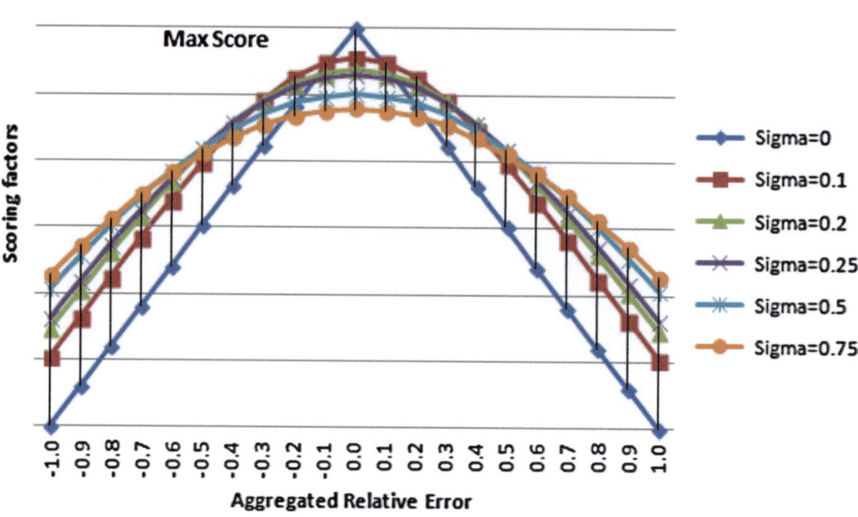

Fig. 5 CRPS scoring mechanism for different errors

predicted period differences. As it can be seen from Fig. 4, there is only one peak, the modified scoring function (Eq. (3)) is also a *strictly proper*.

Figure 5 shows the realization of scoring factors for different errors and confidence level. As pointed out before, the CAs will report their predictions of device usage in the mean of relative error (Eq. (1)) aggregated over all devices. Since, the CAs are aware of the scoring system used by the EPs, they have the liberty to choose associated confidence level (i.e. the sigma; σ). From the graph presented in Fig. 5, it is important to notice that,

a. when a CA is highly confident about its prediction (i.e. $\sigma_u = 0$); highest score is rewarded only when the realized absolute error is zero
b. when the realized error is relatively higher, the CA will be benefitted to report lower confidence (i.e. higher values of σ)
c. CAs do not know the exact shape of the function when it declares the prediction, since the actual error only realized when the event occurs

However, a CA has ideas how it will be scored. For instance, if it is likely to make a larger error, it implicitly chooses the function that will penalize it lower by reporting larger σ. On the other hand, if it is confident of its prediction accuracy, it will report a lower σ. By this way, we ensure CAs to report truthfully about their prediction intentions. Moreover, our model assume no *collusion* between the participating agent devices (CA and EP). As we mentioned (and will show in later section also), the prediction that EP makes regarding the aggregated energy requirement for its CAs, does not involve any imposition or disturbance towards

CAs' device prediction. Rather, EP uses CAs' intentions as a base to report its prediction. Therefore, EP also exhibits the *truthfulness* property. So, the agent devices (CA and EP) used in this model are *truthful* and *non-collusive*.

3   GENCO and EP: Cost Optimization

Based on the device shifting prediction of CAs, EP will try to produce a *potential reward* (pr) which is actually based on the *shifting probability* of a particular device by the amount of shifting. In ideal case, where CA's device commitment prediction coincides with the actual one, there will be no shifting. Taking such scenario in mind, the *shifting probability* (SP) function is chosen to be concave and assumed to be increasing in pr and decreasing in shifting time t. Thus SP is defined as,

$$SP(pr, t) = \frac{pr}{(1 + t)^2} \tag{4}$$

Given the rewards pr_i ($i = 1 \ldots N$) in each period, the amount of demand shifted out of each period i into each period $k \neq i$ is calculated first. The total cost of offering *potential reward* by summing up the demand shifted into period i is calculated as,

$$X_{ep} = \sum_{i=1}^{N} pr_i \sum_{u=1}^{U} \sum_{k \neq i} \sum_{j \in DV_u^k} e_j SP_j(pr_i, |k - i|) \tag{5}$$

where e_j is the amount of energy required by device j. DV_i^u is the set of the devices to be committed to u at period i. EP's cost minimization function for providing *potential rewards* based on the CA's device shifting probabilities. The *potential reward* (pr) is actually given in every hour. Therefore, the reward will be distributed on the consumers who owns those devices at which are committed at that particular hour. EP will be incentivized by GENCO using CRPS function. Which is why, they will report regarding their energy demand (which is in fact the predicted demand summing over all CA) in a form of *Gaussian Distribution*. Total predicted demand for user u is shown in Eq. (8)

$$D_u^p = \sum_{i=1}^{N} \sum_{u=1}^{U} \sum_{j \in DV_u^k} e_j pr_i \tag{6}$$

Therefore, the estimated demand for a particular EP, D_u^p is the summation of individual prediction of all CA. Taking account of this prediction, the EP will report the aggregated *relative error* to GENCO as

$$\sigma_{EP}^2 = \frac{\sum_{u \in EP}(D_u^p * \sigma_u)^2}{(\sum_{u \in EP} D_u^p)^2} \tag{7}$$

Upon receiving the EPs' prediction on demand, GENCO will try to minimize its production cost while satisfying EPs' demand. The GENCOs usually operate multiple plants of different types, e.g. gas, hydroelectric (hydro), renewables and coal [6]. These plants may be categorized as base, intermediate, and peak-load. The base-load plants generally have a higher capital cost but low operating cost, and thus run all of the time (e.g., hydro, nuclear). Intermediate load plants (e.g., coal) have a higher operating cost, and peak load plants (e.g., gas turbines) have the highest operating cost. In any given period, if EPs' demand exceeds the base-load capacity, the generator turns to the intermediate-load plants and then finally, to peak-load plants to generate additional electricity. Under such consideration, c_{i1} denotes the marginal additional cost of using intermediate- rather than base-load plants in period i, and c_{i2} denotes the marginal additional cost of using peak- rather than intermediate-load plants in period i. These marginal costs are instances of the random variables; assuming that their actual values are exogenously determined for use in the EPs' optimization problem. The amount of demand in each period i is

$$Y = D_i - \sum_{EP \in EPS} pr \sum_{u \in EP} D_u^p \tag{8}$$

EPS is the set of EP registered to buy energy from that GENCO. Recall that, D_i is the total demand at period i. EPS is the set of EP registered to buy energy from that GENCO. The cost of meeting consumers demand at period i, therefore, is

$$X_{genco} = \sum_{m=1,2} c_{im} [Y - C_{im}]^+ \tag{9}$$

where C_{i1} and C_{i2} denote the base- and intermediate-load capacities, respectively, for period i. The optimization problems for both EP and GENCO have the same control variables (*potential reward*). Therefore, they can be combined as a single optimization problem. Which is

$$\begin{aligned} \min_{pr} \; & \left[X_{ep} + X_{genco} \right] \\ s.t. \; & pr \geq 0 \end{aligned} \tag{10}$$

The above equation ensures that, GENCO and EPs are able to minimize the cost of satisfying EPs' demand (which is actually aggregated demand from the registered CAs) and the generation cost, respectively.

3.1 Discount Distribution

After calculating the *potential rewards*, they will be distributed among EPs. First, the contribution of each EPs is determined by the fraction of their CRPS score by normalizing the *total potential reward*. We recall that, EP provides its predicted *relative error* as σ_{EP}^2 in the form of Gaussian Distribution and upon realizing actual demand, its actual relative prediction error, say δ_{EP} will be determined. Therefore, the discount EP will receive for their truthful prediction is

$$Discount_{EP} = \frac{\left(\sum_{i=1}^{N} pr_i\right) * CRPS(\sigma_{EP}^2, \delta_{EP})}{\sum_{EP \in EPS} CRPS(\sigma_{EP}^2, \delta_{EP})} \times \log\left(\sum_{u \in EP} D_u^a\right) \qquad (11)$$

The actual realized energy demand, D_u^a for a particular user u, is added in the as a log factor with the discount distribution. This inclusion will encourage elaborate participation of CAs. The more CAs participate, the associated EP will get better discount. Therefore, it is beneficial for an EP to promote all of its CAs to participate in the pricing scheme. The *truthfulness* Eq. (11) will be given in next section.

Now EP's discount should be distributed among the consumers. This discount will be shared based on the scoring rule defined as Eq. (3). Note that, at the time of realization, EP already knows whether that CA is true to her device scheduling prediction. Thus, CRPS is determined according to Eq. (2) where actual weighted average of cumulative error, δ_u. However, as we pointed out, CRPS itself is not good enough to prevent CA to misreport about its true intention. The scoring rule for CA is, therefore, further modified by using Eq. (3). The scoring factor for u at period i is therefore, defined as

$$sf_u^i = \frac{revenue(discount_{EP}) * SR(\sigma_u^2, \delta_u)}{\sum_{u \in EP} SR(\sigma_u^2, \delta_u)} \times pr_i \qquad (12)$$

Note that, the EP does not necessarily always distribute the discount as a whole. Instead, it may choose (in most likely cases) to provide some fraction of it. In this way, EP generates some revenue. Such function is defined as *revenue(discount)*. In this model, we reduce the discount by 40 % to keep the rest as revenue for EP. Finally, the consumer will be rewarded using sf_u^i. sf_u^i is scaled between 0 to 30, since the usual convention is not to offer discount more than 30 %.

4 Algorithm and Properties

The steps each CA takes to report its prediction is shown below. For a given date, CA (which is located in ECC) will follow these steps after it receives the day-ahead pricing signal for the next day.

Step 1. Accumulates the preference of consumer for next day.

Step 2. Integrates the day-ahead price curve.

Step 3. Prepares the operating schedule for each of the smart devices which is registered to ECC.

Step 4. Determines the *prediction accuracy* for each device (based on the historical data for that device) in the form of *Normal Distribution*.

Step 5. Accumulates the *Normally Distributed* prediction accuracies for all devices and build a unified Normal Distribution (zero-ing the mean error) according to the Central Limit Theorem. Save the cumulative standard deviation (σ_u^2) as *confidence*.

Step 6. Reports the device prediction schedule and confidence to associated EP.

The next set of steps will show the process of an EP reporting to GENCO about its prediction about demand

Step 1. Collects all reported predictions and associated confidences from the assigned CAs.

Step 2. Determines the energy requirement for each period by taking account the per period energy demand for all of the active (predictive) devices and confidence of the CAs which report those predictions

Step 3. Since EP has to predict the energy demand, it calculates the error as a Normal Distribution. This distribution will be calculated from the individual prediction *confidences* (which are *standard deviation* actually) of CAs according to Central Limit Theorem. Therefore, EP's *confidence* (σ_{EP}^2) is merely a culmination of CAs' (which were assigned to be served by that EP) prediction.

Step 4. Finally reports the energy prediction and the confidence to GENCO.

Then GENCO process these information from EPs, calculate the potential rewards and distribute them according to the process described in Sect. 3.1. EPs in turn divide the discount among CAs.

4.1 Truthfulness

Truthfulness is an important measure to validate an agent based protocol or strategy. In a competitive environment, where multiple agents are negotiating or bidding over certain resource (*auction mechanism* as such), the designed protocol must consider the interdependency among bidders. That being said, the payment or utility (or discount in our context) of one agent depends on the bidding (or reporting) of other agents. However, recall the fact that, in our model, the agents are not competing. Therefore, the definition of *Truthfulness* is not exactly same here as defined in *auction mechanism* design. Rather, this model in particular, we can think of each agent independently reports its confidence as *truthfully* as possible to get the highest discount. The mechanism here (*scoring rule*) designs the discount such a way that, it will elicit the private information of agents truthfully. Assuming the agents here are *rational*, they always go for the report which essentially rewards them with the highest discount possible. In other words, if an agent cheats (or being manipulative) while reporting, it will always end up not getting highest discount. On other note, we like to mention that, the model itself contains some uncertainty which the agents predict while the designed *scoringrule* does not posses. Now, one obvious way to deal with *truthfulness* is to set a mediator which will observe the behavior of all participated agents and thus ensures all agents report truthfully. However, mediator based protocols impose several difficulties such as centralizations, private information elicitation, etc. At the same time, as the number of agents grows higher, it will become difficult for a single mediator to control and process the information. Therefore, the best way to ensure is to design the protocol such a way that, its always the *best interest* for an agent to report truthfully in order to get maximum expected utility. In the game theory term, we can call it agents *best response*. However, since these agents here are not competing, we are not interested to find the *Nash Equilibrium*. Nonetheless, ensuring *best response* as the *best strategy* for an agent is central to the success of the designed scoring rule. Recall that, CRPS (Eq. 2) is a *strictly proper* scoring rule. The proof of strict propriety can be found in [4]. *Strictly proper* scoring rule inherently ensures the agents to be truthful when CRPS is used as the discount scheme. Since, it has a single optimum, declaring true intention as prediction always maximizes the expected utility of reporting agent. As shown in Eq. (3), we have redefined the scoring rule by extending CRPS. So, we have to make sure, the redefined function still holds the *truthfulness* propriety. The part, $\left[\min_{d \in DV_u} \frac{4+(P_d^a - P_d^p)}{e^{4(P_d^a - P_d^p)}} \right]$ in Eq. (3) actually produces a value which makes the equation just an affine transformation of CRPS (i.e. sign of the derivatives, the gradient and concavity are same as the original form of function). Therefore, it is evident that, the scoring rule defined in Eq. (3) is also *truthful*. At the same time, the discount function of CA defined in Eq. (12) also reflects the truthful distribution of $discount_{EP}$ according to the scoring rule. For EP agent, the strictness propriety of CRPS sustains. The reason is that, EP does not actually imposing some non-linear transformation of CRPS. Rather, according to Eq. (9), it just accumulating the predictions of registered CA (σ_u) and reports the distribution of GENCO. Therefore, it can safely be said that, if CAs reports are *truthful*, so is the EP's report.

Proposition 1. *The optimization problem defined in Eq. (10) is a convex optimization problem. We can say that by showing the fact that, the shifting probability function is concave given the potentialreward, pr increases as the shifting error, t decreases. From the definition of SP (in Eq. 4), SP is maximum (i.e. full potential reward, pr) when t is zero and thus monotonically strictly decreasing. Therefore, SP is a concave function. This statement can be further strengthen by showing that the second derivative of Eq. (5) is positive-definite in the range of feasible pr. In other words, the marginal cost of discount (potential reward, pr) offering is lower than the discount itself.*

5 Agent Simulation

In this section, some data analysis and simulation results are presented in order to verify the feasibility of the hierarchical scoring rule based pricing mechanism as well as to demonstrate the ability to reduce GENCO's cost of electric generation and flatten electric usage over some periods. The following cases are considered

Case 1. Simple case of 1-GENCO, 1-EP and 1-CA
Case 2. Extended case of 1-GENCO, 10-EPs and 100-CAs

The real parameters are taken based on Ontario Independent Electric Operator [10]. The base-load plant in Ontario is hydroelectric; (assume 60 % are base-load plants). The production capacity of each plant is taken as constant across different periods of a day for the purposes of simulation. The intermediate-load consists of coal (operating at 20 % efficiency, as is consistent with IESO) and the remaining hydroelectric plants. Finally, the peak plants are gas turbines, which are the most expensive to operate. The slopes of the cost functions for base-, intermediate- and peak-load plants are taken from the production estimates in [9]. The marginal costs of moving from intermediate- to peak load and base- to intermediate-load plants are calculated to be \$62.46/MWh and \$18.54/MWh respectively. Since its a GW based power system, the system data is effectively scaled down to provide simpler simulation conditions. The prices are also equivalently scaled into kWh level.

5.1 Case 1

Assuming the electricity usage of a single consumer, Fig. 6 presents the periodically expected device scheduling after applying the smart pricing scheme. While the quantitative results of these simulations will vary from market to market, the qualitative results suggest that smart pricing can indeed help GENCOs and EPs to even out consumption over the day and reduce the energy requirements from peak-load plants. Scrutinizing the cyclic electricity consumption pattern, it can be shown that for four consecutive days of energy consumption of 1 consumer reported is 142.1 kWh. However, before using the reward based pricing scheme, the total consumption recorded for a single consumer was 170.35 kWh (based on the single household consumption determined according to the data presented in [10]). Therefore, for a single consumer, the proposed scoring rule based reward scheme can reduce energy consumption down to 20 %.

Figure 7 shows the discount corresponding to the same energy consumption pattern as discussed in previous paragraph. It is noted that the discounts are roughly cyclical, as might be expected. If we check the pattern, it is clearly seen that, in case of peak demand hour, the reward is minimum which states the fact that, it becomes

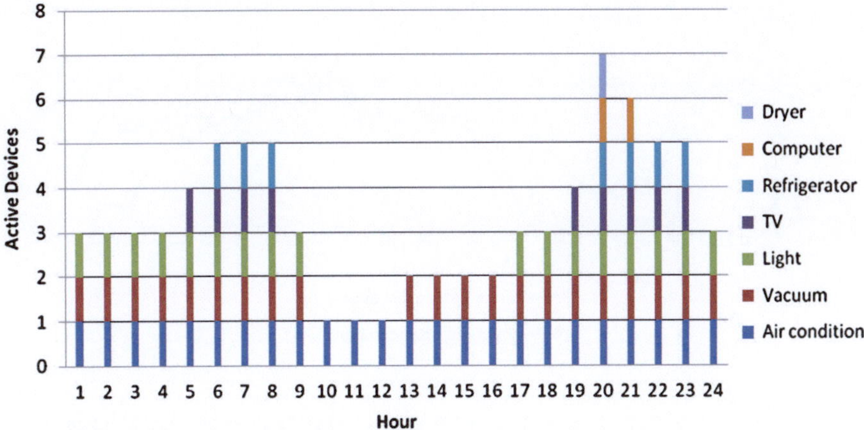

Fig. 6 Hourly active devices for a single consumer in a 24-h period after running smart pricing scheme

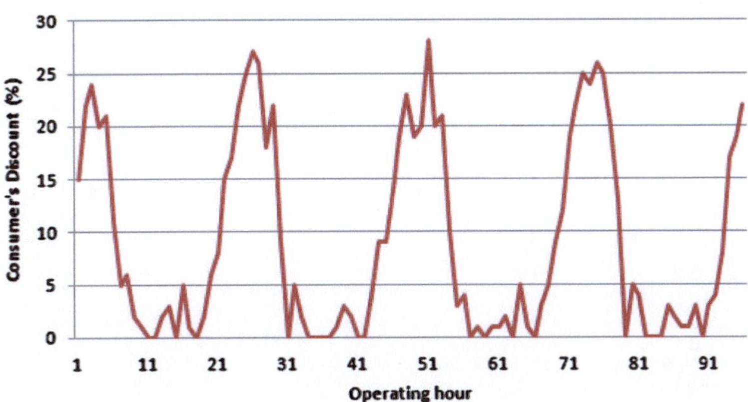

Fig. 7 Discount provided to a single consumer for 96 h. Total energy consumed 142.1 kWh

difficult to make an accurate prediction in peak hour. Figure 8 depicts the effect of smart pricing scheme on pricing. We can see that, a CA can effectively reduce payment towards EP if it truthfully reports about its device scheduling on the basis of the day-ahead price signal and shifts them in lower priced period. In order to portray the effect of accuracy factors of prediction (the reported σ by CA), we have showed two separate CAs and their payments in per kWh. The effect is shown in Fig. 9. In this figure, consumer (CA) 1 predicts accurately and relatively truthful than consumer (CA) 2. Therefore, CA1 will pay less than CA2.

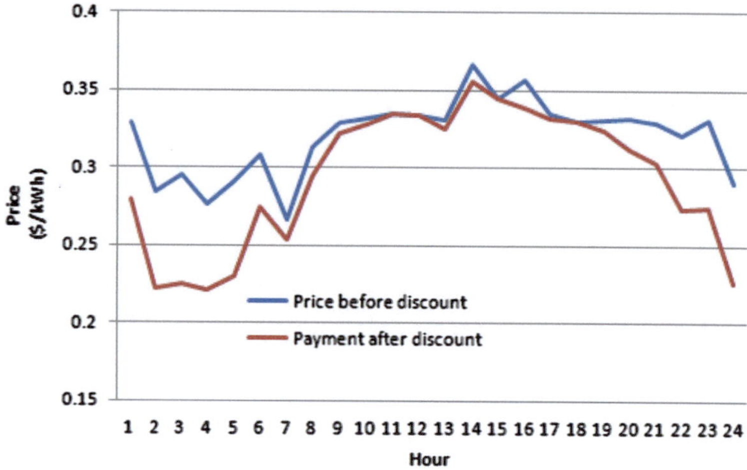

Fig. 8 The price curve Vs. the payment which occurred before and after applying smart pricing scheme. This price signal is taken from 1st March, 2010

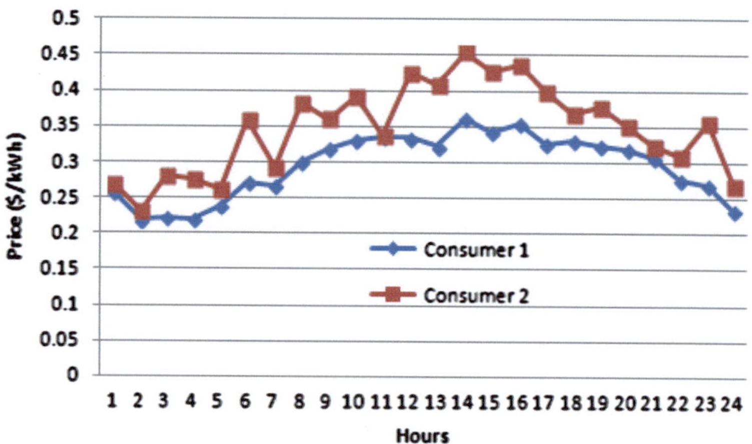

Fig. 9 Effect of prediction accuracy between two consumer agents

5.2 Case 2

To provide the scalability of the proposed method, Fig. 10 is presented. It shows effect aggregated over 1000 consumers (1-GENCO, 10-EPs and 100-consumers). It is noted that the peak-to-average (P2A) ratio of the consumption pattern before using scoring rule based smart pricing is 2.55 while it comes down to 1.91 when using the proposed pricing scheme. Qualitatively speaking, the P2A ratio is down by approximately 25 %. Therefore, the proposed scheme works better when the

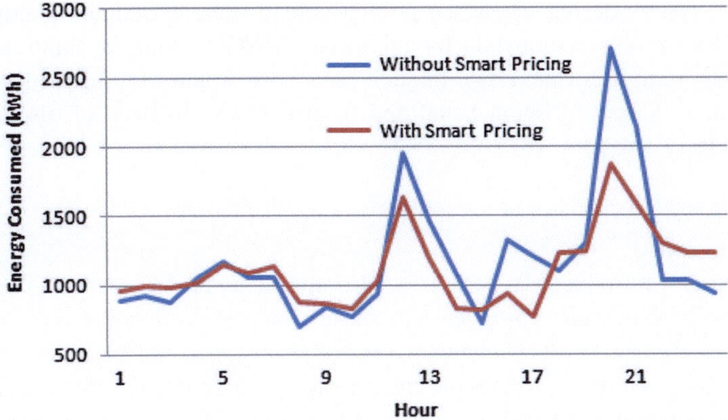

Fig. 10 Energy usage before and after using of scoring rule based smart pricing; aggregated over 1000 consumers

number of consumers is higher. So, it can be said the method is practically viable and scalable. Moreover, such measure reflects the fact that the strictly proper scoring rule based reward mechanism is able to flatten the load demand.

6 Related Works

The earlier works that explored the potentials of such pricing schemes are mostly from the perspective of consumers on the light of scheduling devices according to the prediction of future prices. As for an instance, [8] proposes a mechanism for predicting prices one or two days in advance. The household devices/loads can be scheduled as to balance impatience with the will to save money. Here several users share a power source and simultaneously scheduling energy consumption in a distributed manner. EP's problem of determining prices based on the consumer response has been studied in several literatures. For example, [3] uses real data to quantitatively forecast consumers' scheduling of energy consumption. Another work reported in [2] presents a day ahead pricing model for smart houses considering providers' profit maximization, energy consumption minimization using a decomposition strategy. Although the other works do deal with aggregated demand across the consumers at different times, none of these ever take heterogeneity at the device level into account.

Additionally (and importantly), none of these works ever considered properly incentivizing the consumers for accurately scheduling their devices for future periods as well as making quality estimation of device scheduling for coming periods. Consumers' reactions on dynamic pricing effect the scheduling of the household devices. Since there is no standard device commitment mechanism exists, the consumer may misreport their scheduling preferences which leads to an inefficient system design where the pricing is tend to be biased and manipulative. Again, EPs are also responsible for reporting GENCO about the total amount of energy they require for next day. In such case, correct and truthful prediction also provide them a way of being benefitted from GENCO in form of discount over buying price.

> **Conclusion**
>
> This paper introduces a new smart pricing scheme considering a model consists of generators, provider and consumers. The formulations are carried by devising a truthful mechanism for both consumer and provider entities where they will report their true intentions regarding device scheduling and energy demand, respectively. The scoring system (facilitating *Continuous Ranked Probability Score*) is designed such a way that, it will force consumer agents to report their true beliefs towards providers. On the other hand, provider agents themselves are incentivized to report the energy demand to generation companies (GENCO) truthfully to get a discount over price. The conducted simulation results show that, the proposed smart pricing scheme is able to reduce the total energy consumption as well as consumers payment towards providers. Therefore, consumers are benefitted since they paid less than the actual price and we have a cleaner environment with reduced energy production. As a future study, we will try to model the device sensitiveness towards scheduling and apply such mechanism for higher scaled smart power system.

References

1. Brier, G.W.: Verification of forecasts expressed in terms of probability. Mon. Weather Rev. 1–3 (1950)
2. Doostizadeh, M., Ghasemi, H.: A day-ahead electricity pricing model based on smart metering and demand-side management. Energy **46**(1), 221–230 (2012)
3. Faruqui, A., Wood, L.: Quantifying the benefits of dynamic pricing in the mass market. Edison Electric Institute, Tech. Rep (2009)
4. Gneiting, T., Raftery, A.: Strictly proper scoring rules, prediction and estimation. J. Am. Stat. Assoc. **102**(477), 359378 (2007)
5. Harry, R., Alex, R., Gerding, E.H.: A scoring rule-based mechanism for aggregate demand prediction in the smart grid. In: AAMAS '12: Proceedings of the 11th International Conference on Autonomous Agents and Multiagent Systems (2012)
6. Masters, G.: Renewable and Efficient Electric Power Systems. Wiley Online Library (2004)
7. Matheson, J.E., Winkler, R.L.: Scoring rules for continuous probability distributions. Manag. Sci. **22**, 10871095 (1976)
8. Mohsenian-Rad, A.H., Wong, V.W.S., Jatskevich, J., Schober, R.: Optimal and autonomous incentive-based energy consumption scheduling algorithm for smart grid. In: Innovative Smart Grid Technologies, pp. 1–6. IEEE (2010)

9. Morgan, J.: Comparing energy costs of nuclear, coal, gas, wind and solar. http://nuclearfissionary.com/2010/04/02/comparing-energycosts-of-nuclear-coal-gas-wind-and-solar/ pp. 155–212 (2010)
10. Ontario, I.: Market data. http://www.ieso.ca/imoweb/marketdata/marketData.asp pp. 155–212 (2011)
11. Robu, V., Kota, R., Chalkiadakis, G., Rogers, A., Jennings, N.R.: Cooperative virtual power plant formation using scoring rules. In: Twenty-Sixth Conference on Artificial Intelligence. AAAI (2012)
12. Samadi, P., Mohsenian-Rad, A., Schober, R., Wong, V.W.S., Jatskevich, J.: optimal real-time pricing algorithm based on utility maximization for smart grid. In: Conf. on Smart Grid Communications. IEEE (2010)
13. Xia, L., Conitzer, V.: Finite local consistency characterizes generalized scoring rules. In: International Joint Conference on Artificial Intelligence (2009)

Evaluation of Route Assignment Method with Anticipatory Stigmergy Under Distributed Processing Environment

Jun Takahashi, Ryo Kanamori, and Takayuki Ito

Abstract In this paper, we propose a foresight route providing method based on anticipatory stigmergy for collecting near-future traffic position. A foresight route providing method combine the previous method that utilize past travel time of probe vehicle due to efficiency of traffic flow. In this model, all probe vehicle submit their near-future traffic position as anticipatory stigmergies and are allocated among foresight route method and previous method based on the allocation ratio. Moreover, collecting dynamic traffic conditions in distributed processing environment is introduced as implementation environment with reality. A distributed processing environment are collected links information limited by the area. This study considers the efficiency of combination with past travel time for a few minutes and near-future traffic position in both a distributed and centralized processing environment. That information is collected in all links.

Keywords Distributed processing environment • Stigmergy • Traffic information • Traffic simulation

1 Introduction

Vehicle became indispensable to the daily life of many people. There are many problems to be solved decline of economical efficiency due to congestion, traffic congestion, deterioration of environmental such as global warming and air pollution, and traffic accident. In addition to the promotion of next-generation vehicle such as electric vehicle, traffic operation and management by ITS (Intelligent Transport System: Intelligent I due to the transportation system) like signal control in response to traffic conditions and road pricing have lately attracted attention as a method of solving these problems. Among them, providing route and traffic information with the spread of the navigation system and vehicle sensor, the technology becomes also more sophisticated. Recently, it can be observed the speed and position every

J. Takahashi • R. Kanamori • T. Ito (✉)
Nagoya Institute of Technology, Nagoya, Japan
e-mail: takahashi.jun@itolab.nitech.ac.jp; kanamori.ryo@nagoya-u.jp; ito.takayuki@nitech.ac.jp

© Springer Japan 2015

Q. Bai et al. (eds.), *Smart Modeling and Simulation for Complex Systems*, Studies in Computational Intelligence 564, DOI 10.1007/978-4-431-55209-3_9

moment as a sensor of the vehicle and chased the change of traffic in dynamically. Moreover a route providing method to utilize probe vehicle information based on the pattern of past travel time is suggested [9]. In addition, centralized traffic conditions of probe vehicle collected by private companies that were acquired after the Great East Japan Earthquake, was published and updated on the map.

More sophisticated coordination methods are becoming feasible by utilizing the current traffic information. More precise traffic information can be provided by car navigation systems with GPS and probe-vehicle information. These data are stored in central servers as **long-term** memory that can provide stochastic traffic congestion information to vehicles. Such information technologies have been already applied in the real world. Researches in the field of transportation and multi-agent systems have been focusing on dynamic **short-term** memory. Vehicles share this dynamic information, and drivers can choose their routes more dynamically based on real-time information. This short-term traffic information is usually modeled as a stigmergy. **Stigmergy** [4] has been used for indirect communication for cooperation among agents. For example, ants' pheromone is a kind of stigmergy for cooperation among them. In this case, ants are modeled as agents in multi-agent models and also as vehicles in traffic situations. Vehicles can estimate their nearest future situation based these stigmergies.

In this paper, we propose an anticipatory stigmergy model for decentralized traffic congestion management. Recently, there are several studies and practices for observing traffic flow and providing information on traffic congestion. These are usually done by counting the vehicles that pass particular locations using sensing gates that are usually placed on the arterial roads. Such information is broadcasted as current information to vehicles. It is rarely stored and cannot work as shared memory. One drawback of these stigmergies approaches is that handling **near-future** congestion remains problematic because stigmergies are basically **past** information. In this paper, we propose **anticipatory** stigmergy for sharing information about **near-future** traffic. In our model, each vehicle submits its near-future **intention** as anticipatory stigmergy and reschedules its route plan based on the submitted anticipatory stigmergies and is assigned to the routes based on actual past travel time and anticipatory stigmergy in order to avoid congested.

However, it is difficult to observe of the speed and the position information provided by vehicles equipped GPS and the search and provide for the route in wide area real time with the spread of vehicles recently. Especially, the limited range of dynamic travel time for a few minutes and near-future traffic handled as a stimgergy that improves cooperation indirectly is realistic. In this paper, a limit at a distance is introduced in order to implement a distributed processing environment. Concretely, link evaluation value is calculated from dynamic traffic conditions based on our proposed method of this study if link in the range can be receive dynamic travel time of a few minute past and estimated traffic volume of a few minutes later. Otherwise, link evaluation value is set the stored data of past travel time that can store in advance. we investigate the verification of efficiency of our proposed method in both distributed and centralized processing environment at test road network.

2   Related Works

Many approaches have handled traffic congestion from real-world approaches to theoretical ones. Distributed approaches have been widely studied in the field of multi-agent systems and artificial intelligences. A very large-scale survey paper [1] investigated the entire transportation field with multi-agent systems from traffic congestion to railway transportation. Some work is focusing on Ant Colony Optimization (ACO) [5, 6], which can be fit into congestion avoidance problems. We also believe that the swarm intelligence approach can be adapted for traffic congestion control. Reference [2] proposed anticipatory vehicle routing using multi-agent systems. Here, their multi-agent systems actually resemble ACOs. For anticipation purposes, they use ACO as an optimization mechanism before the actual routing. Reference [10] models a vehicle as an ant in an approach that closely resembles ours. The difference is that reference [10] adopts only pheromone mechanisms without probe information used in the real world. Their methodology could be improved by integrating it with current real-world approaches to reduce traffic congestion. Recently, the paper [3] uses a *inverse* ant colony optimization. It is inverse in the sense that a strong pheromone trace will influence following cars not to follow their predecessors but instead to avoid this road, taking a different route to their goals. Reference [9] provides knowledge on drivers' dynamic route choice behavior using probe-vehicle data. Modeling route choice from real probe-vehicle data is essential because real route choices could be biased by habitual activities. References [11, 13] also approach drivers' dynamic routing modeling. [14] decide use of road resources based on market-based approach. In the paper, these resources are needed reservations. The paper [7] has shown, as a preliminary result, the comparison among some types of stigmergies in transportation management. The authors show the usefulness of the prospective route information to avoid in advance handling, congested as anticipatory stigmergies, the need for allocation in the traffic situation in the near future by gathering the count of vehicles after a few minutes, and analyzes the impact of such traffic capacity reduction from construction for a fixed period of time [7, 8]. But it doesn't consider providing information in distributed environment.

3   How to Collect and Provide Traffic Information

3.1   Case 0: No Information

No traffic information is gathered or provided. Each vehicle finds the best path by Dijkstra search before it departs. We assume several different starting and end points since people have different origins and destinations. The cost of link l can be shown in Eq. (1):

$$v_0 = t_0(l) = int\left(\frac{|l|}{v_{max}(l)}\right), \tag{1}$$

where $t_0(l)$ defines a free flow travel time, and $v_{max}(l)$ defines the maximum speed and $|l|$ is distance of link l.

3.2 Case 1: Combined Long- and Short-Term Stigmergy

A road (link) stores and manages long-term stigmergy (historical) information forever. As long-term stigmergy information, the each link stores the travel time from the vehicles equipped with GPS (i.e., probe cars), and provides them a long-term stigmergy value $v_l = ave + sd * 0.5$, where ave is the average and sd is the standard deviation of all stored data of each link. Each probe vehicle utilizes this long-term stigmergy information to make a new plan by Dijkstra search before its department. Long-term stigmergy is updated everyday, where in our simulations x is 1 (i.e., daily update).

A road (link) stores and manages stigmergy information for only the most recent one unit-times (minutes). As short-term stigmergy information, the each link keeps storing data about the travel time of probe cars for only the most recent one minutes, and provides short-term stigmergy value $v_s = ave$, where ave is the average. the standard deviation is not utilized because enough data is not collected in the most recent one unit-times. If there is no vehicle passing the link for 1-min, short-term stigmergy value is $v_s = v_0$. Each probe vehicle utilizes this short-term stigmergy information to search new route by Dijkstra algorithm every one unit-times (minutes).

In this paper, distributed processing environment and to limit the providing information of dynamic traffic conditions in the environment are assumed. The range can be collected on the short-term link evaluation value(v_s) is up to two destination nodes as seen from the prove vehicle, the link which cannot be collected the short-term link evaluation value(v_s) utilize the stored data of past travel time (long-term stigmergies).

A road (link) stores and manages long- and short-term stigmergies. As mentioned before, the long-term stigmergy information is the value of v_l that is updated daily, the short-term stigmergy information is the value of v_s that is updated every one unit-times (minutes). Each probe vehicle utilizes combined long- and short-term stigmergy information to make a new route plan by Dijkstra search every five minutes. Equation (2) shows how to combine long- and short-term stigmergies, and v_{ls} is the combined stigmergy information:

$$v_{ls} = v_s * (1 - w) + v_l * w, \tag{2}$$

where v_s is the short-term stigmergy value, v_l is the long-term stigmergy value, and w is the weight of the long-term stigmergy ($0 \leq w \leq 1$).

3.3 Case 2: Anticipatory Stigmergy

Every five unit-times (minutes), all probe vehicles find the best route to their destination node based on long-and short-term stigmergy, as in case 2. Here, they submit (as a link) where they will be in the next one unit-times (minutes). This is how we define anticipatory stigmergy. Then they can confirm the traffic situation in future and search the best route based on the anticipatory stigmergies. Equation (3) shows the heuristic cost of link l by using anticipatory stigmergies, which are average travel time calculated by link performance function defined by The Bureau of Public Road (BPR) in U.S. [12]:

$$v_a = t_0(l)(1.0 + \alpha * (\frac{Vol(l)}{Cap(l) * 0.4})^\beta).$$ (3)

$Vol(l)$ is the total number of probe vehicles in near-future on link l gathered by anticipatory stigmergy. Function $t_0(l)$ is a free flow travel time, and function $Cap(l)$ is a capacity of link l that is adjusted adequately (in this study for the cellular automaton model, the adjustment value is set 0.4 because the condition to drive freely is a half of capacity). $\alpha = 0.48$, and $\beta = 2.48$. This cost function v_a is a heuristic; if there are many vehicles, then v_a will be increased briefly. However, in the case of provision of dynamic traffic conditions in distributed processing environment, the range can be collected on information of anticipatory stigmergy is up to two destination nodes as seen from the prove vehicle, the link which cannot be collected the information utilize the stored data of past travel time (long-term stigmergies). Therefore, the link travel time in distributed processing environment is Eq. (4).

$$\begin{cases} v_a = t_0(l)(1.0 + \alpha(\frac{vol(l)}{\gamma \times Cap(l)})^\beta), & \in \text{area of dynamic traffic conditions} \\ v_a = v_l(l), & \notin \text{area of dynamic traffic conditions} \end{cases}$$ (4)

Every 60 s, all probe vehicles search the best route to their destination node based on a link travel time with anticipatory stigmergies (Eqs. 3, 4). In this case, we introduce a strategy to assign probe vehicles reasonably into the two routes, that one is a route searched with combined long- and short-term stigmergies in case 1 and the other is a route searched with near-future information. Although there are various criteria of assignment, in this study, a rest of straight-line distance to the destination is adopted. A concrete procedure to assign probe vehicles into two routes is as follows.

- If the number of probe vehicles on the link searched with the traffic information of anticipatory stigmergies is larger than the congestion level, which is a half of capacity, probe vehicles are sorted randomly.
- In the concentrated situation, the upper 50 % these vehicles are assigned to the link that is recommended as the best route with anticipatory stigmergies as shown in case 2, and the rest of drivers are assigned to the link that is searched with combined long- and short-term stigmergies as shown in case 1. Otherwise, all drivers choose the link on a route calculated in case 1.

4 Traffic Simulator

4.1 Cellular Automaton Model

In order to treat each vehicle as a discrete one (not continuous), our developed traffic simulation model is one of a cellular automaton model. This simple rule is famous as a "Rule 184" [15]. A concrete flow of the vehicles is as follows.

- Each link corresponding to the road section between intersections sets the distance and regular speed. These links are divided in the number of cell which vehicles can move at one time step. The number of cell is equal to free travel time (= distance of link/free flow time). In this paper one time step is one second.
- It can be one vehicle enters the cell, and each vehicle determines whether the movement in accordance with the status of the vehicle in the previous cell each time step.
- If the vehicle is present in the previous cell, it can not be moved in the time step, remain in the cell in the presence.
- Otherwise, If the vehicle does not exist in the previous cell, the vehicle can move forward cell in the time step. However, vehicles can move only the straight direction, can not change lane. Traffic simulator of this study is the main objective comparison of indicators including the travel time providing route information to avoid congested. These rules are very simple because it is not intended to reproduce the travel time and actual vehicle behavior. Elaboration of traffic simulator is a subject of the future.

4.2 Road Network

We model a road network as a graph. Let directed graph $G = (N, E, Cap, t_0)$ serve as a model of the road network, where N is a finite set of nodes that model intersections, and E is a set of links that model one-way roads among intersections. Link $l = (n, n')inE$ if and only if there is a link that permits traffic to flow from intersections n to n'. Function $Cap(l)$ defines the capacity on link l. Function defines a free flow travel time of link l. Each vehicle i has origin node and destination node . $|l|$ is the length of link l. We assume two road classifications: arterial and ordinary. Arterial roads have two lanes while ordinary roads have one lane. The following is the procedure to determine the characteristic values for each link in this paper.

- The length of link is uniform 200 m.
- The link in road network are classified into an arterial road or an ordinary road.
- If a link is an arterial road, the number of lanes of link l is two. Otherwise, it is one.

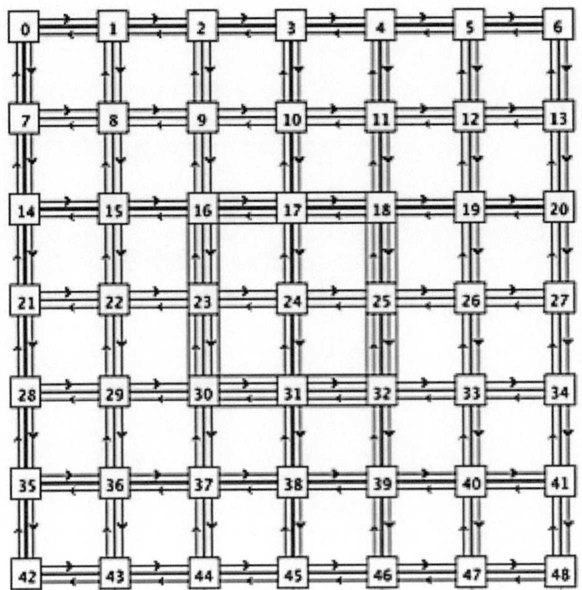

Fig. 1 Network

- If a link is an arterial road, the maximum speed of link l, $v_{max}(l)$, is sampled from *Uniform*$(20, 30)$ km/h. Otherwise, it is sampled from *Uniform*$(15, 25)$ km/h.
- The capacity *Cap*(l) is calculated from the number of cells and lanes.

In this study, we use a simple road network (Fig. 1), where $16\longleftrightarrow 17$, $17\longleftrightarrow 18$, $16\longleftrightarrow 23$, $23\longleftrightarrow 30$, $18\longleftrightarrow 25$, $25\longleftrightarrow 32$, $30\longleftrightarrow 31$, $31\longleftrightarrow 32$ are set arterial roads (the number of lane is two), and the others are the ordinary roads (the number of lane is one).

4.3 Dynamic Traffic Conditions Collecting Range

Figure 2 shows that the range which each vehicle enable to receive dynamic traffic conditions. The range is two destination link from nodes which vehicle exists. In Fig. 2, these link is painted black. Dynamic traffic conditions indicate stigmergy information for only the most recent one unit-times (minutes) (short-term stigmergy) and for near-future (anticipatory stigmergy). In addition, the node which the vehicle exists indicate the traveling direction side of the link that the vehicle is passing.

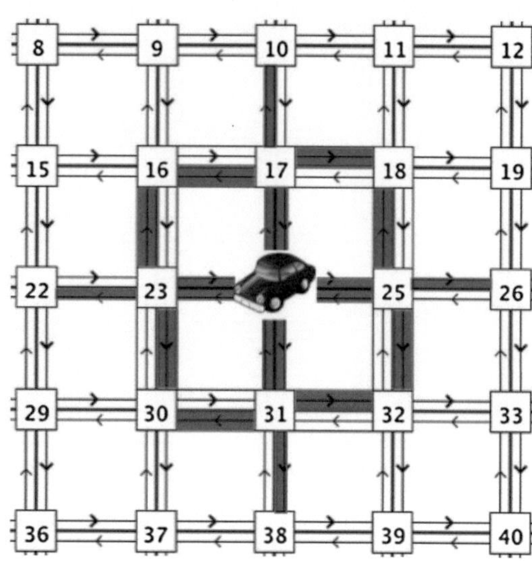

Fig. 2 Available receive dynamic data

4.4 Origin-Destination Traffic Volume

OD(origin-destination) traffic volumes are 800 vehicles. The 200 vehicles start from node 0 to node 48(0—→48). Another 200 vehicles start node 2 to node 45(2—→45). Another 200 vehicles start node 4 to node 45(4—→45). The other 200 vehicles start node 6 to node 42(6—→42). This setting is considering the diversity of route choice and generation of congestion. We assume that 75 % vehicles in each OD pair have a device to send and receive information (i.e., probe vehicle like car-navigation systems that can handle stigmergies by GPS and so on).

5 Experimental Results

In this paper, we compare the effective of traffic flow and verification of allocation strategy in all case (case 0–2). The feature of case is as following.

- Case0 : No Information
- Case1 : Combined Long- and Short-Term Stigmergy
- Case2 : Anticipatory Stigmergy without Allocation Strategy
- [Distributed] : enable to collect limited dynamic traffic condition
- [Centralized] : enable to collect all dynamic traffic condition

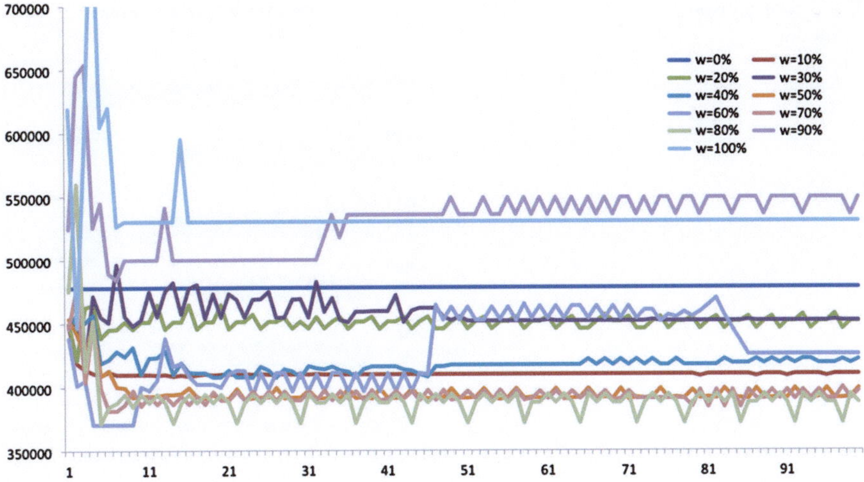

Fig. 3 Sensitivity analysis of Case1 in centralized (100 iteration)

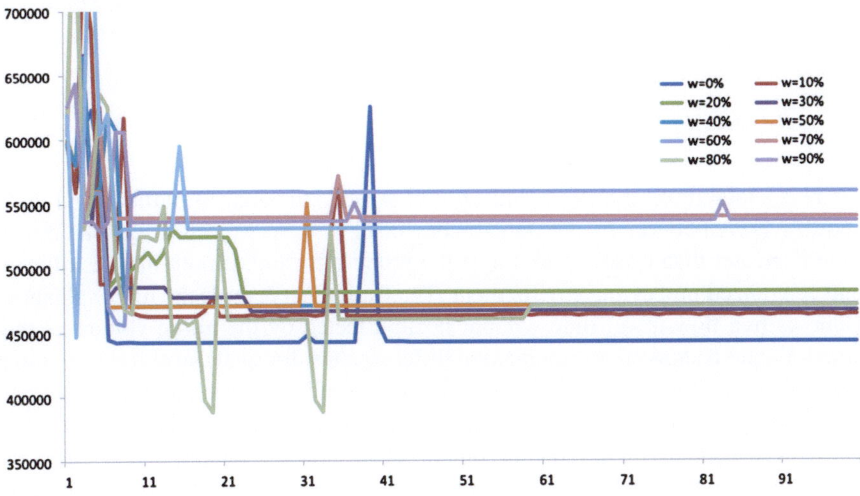

Fig. 4 Sensitivity analysis of Case1 in distributed (100 iteration)

In case 1 and case 2, the evaluation in distributed processing environment and centralized processing environment is conducted. The number of simulations are conducted 100 days. We use five pattern uniform random numbers because the probability of request route negotiation. In this analysis, the result uses last fifty days' result (Figs. 3 and 4).

Fig. 5 Comparison of total travel time

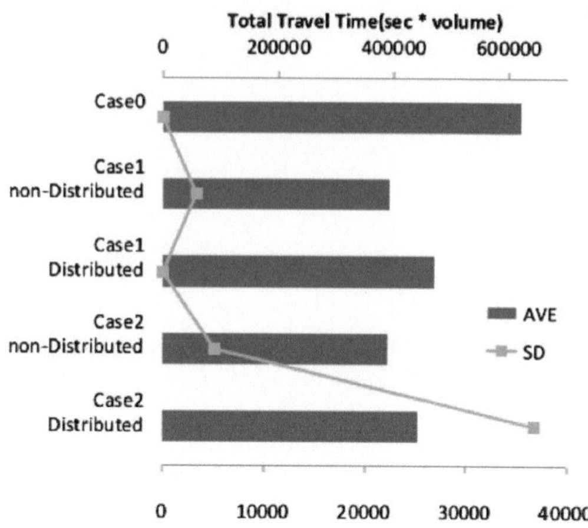

Fig. 6 Signification difference test

	Distributed	t-value
Case1-Case2	×	4.842
(Total Travel Time)	o	5.378
Case1-Case2		6.778
(Time Loss in Congestion)	o	1.050

5.1 Results of Total Travel Time

There is a total travel time as an indicator of typical efficiency of traffic flow, Fig. 5 shows average between 51 day and 100 day in all case. As expected, the total travel time is greater than Case1, 2 to utilize the probe vehicle data both in a distributed and centralized processing environment. The Case0 that regards the link evaluation value as free flow travel time without begin affected by the traffic volume is the worst. Figure 6 shows that signification difference test for total travel time and time loss in congestion both in a distributed and centralized processing environment. Case2 to utilize anticipatory stigmergies in a centralized processing environment is the best. In this evaluation experiment, Case2 is more effective than Case1 to utilize the probe vehicle data of the service level of the present navigation system. Signification difference is confirmed for 5 % level of significance in welch's t test.

Moreover, Figs. 3 and 4 show total travel time difference both a distributed and centralized processing environment. These figure confirms the converge both a distributed and centralized processing environment by repeating simulations. In Case1 of centralized processing environment, hunting phenomenon can be seen than the distributed processing environment because the link evaluation value which link cannot be collected in distributed processing environment is small and the link is selected as route.

Fig. 7 Comparison of time loss in congestion

	Time Loss in Congestion	
	Average [sec]	Standard Deviation [sec]
Case 0	309.4	188.7
Case 1	89.4	47.8
Case 1(Distributed)	83.1	45.8
Case 2	39.9	19.3
Case 2(Distributed)	92.8	47.3

Reduction of time loss in congestion that leads to reduction of travel time, psychological load reduction effect and reduction of traffic accidents is one of the important indicator. A time stayed in congestion from departure is calculated for each driver based on Eq. (5).

$$t_{congestion} = \Sigma(t_{travel}(l) - t_0(l)) \tag{5}$$

where $t_{congestion}$ is a time involved in congestion from departure, $t_{travel}(l)$ is the vehicle's travel times on link l, and $t_0(l)$ is a free flow travel time calculated in Eq. (1) In this preliminary, the weight for long-term stigmergy is 50 % and the route allocation ratio is also 50 % because of preliminary comparison.

Figure 7 shows the averages and standard deviations of time loss in congestion. Coefficient of variation indicates dispersion of time loss in congestion through iteration. At the average of the time loss in congestion, case 2 to utilize anticipatory stigmergy in centralized processing environment is the smallest and case0 not to reflect traffic conditions is the largest. Moreover, at the average of the time loss in congestion in a centralized processing environment Case 2 is smaller than Case 1. Figure 6 shows t-value is 6.778 and statistically-significant difference is confirmed. Thus, It can be seen that time loss in congestion depends on the traffic situation and the service providing qualitative level is improved by allocation. However, in a distributed processing environment, the average of time loss in congestion of Case 1 is smaller than its Case 2. This suggest that in the indicator of total travel time, efficiency is confirmed, but in the indicator of time loss in congestion, the keeping the traffic situation in a wide range is needed.

5.2 Sensitivity Analysis of Route Allocation Ratio

Figure 8 shows sensitivity analysis of route allocation ratio in Case 2 to utilize anticipatory stigmergy for congested link. In this preliminary, 50 % route allocation ratio that the vehicles are assigned up to 50 % of traffic volume of link is most efficient. This suggests the route allocation of this study is needed.

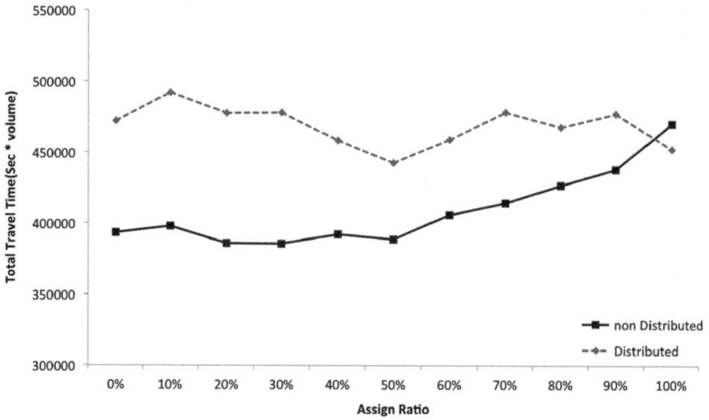

Fig. 8 Sensitivity analysis of route allocation ratio

Therefore, the result of comparison of total travel time shows a foresight route providing method based on anticipatory stigmergy that is shared in near-future position is more effective than in the past method.

Conclusion

In this paper, we analyze about the route information providing method by sending and receiving dynamic traffic conditions in a distributed processing environment. Our preliminary result demonstrated that allocation of the routes based on near-future vehicle position (anticipatory stigmergy) and actual past travel time (long- and short-term stigmergy) in both a distributed and centralized processing environment has statistically significant difference and efficiency. Moreover, at total travel time, half allocation of vehicle($Cap(l) \times 0.5$) in a distributed and centralized processing environment also is efficient in traffic flow. Therefore allocation of vehicle is needed.

In future work, we confirm the verification of route information providing method based on anticipatory stigmergy from the present state to implement elaboration of traffic simulation and large-scale road network reflected a distributed processing environment in preliminary experiment. Besides, foresight providing method based on anticipatory stigmergy in distributed processing environment is not confirmed the signification difference of the indicator of time loss in congestion. We will confirm the improvement of time loss in congestion if the range of collecting dynamic traffic condition becomes wide.

Acknowledgements This study is partially supported by the Funding Program for Next Generation World-Leading Researchers (NEXT Program) of the Japan Cabinet Office.

References

1. Chen, B., Cheng, H.H.: A review of the applications of agent technology in traffic and transportation systems. IEEE Trans. Intell. Transport. Syst. (2010)
2. Claes, R., Holvoet, T., Weyns, D.: A decentralized approach for anticipatory vehicle routing using delegate multiagent systems. IEEE Trans. Intell. Transport. Syst. **12**(2), 364–373 (2011)
3. Dallmeyer, J., Schumann, R., Lattner, A.D., Timm, I.J.: Don't go with the ant flow: Ant-inspired traffic routing in urban environments. In: In the Proceedings of the 8th Workshop on Agents in Traffic and Transportation (ATT2012) (2012)
4. Dorigo, M., Gambardella, L.M.: Ant Colony System: A Cooperative Learning Approach to the Traveling Salesman Problem. IEEE Trans. Evol. Comput. (1997)
5. Dorigo, M., Stützle, T.: Ant Colony Optimization. MIT Press, Cambridge (2004)
6. Dorigo, M., Stützle, T.: Ant Colony Optimization: Overview and Recent Advances. Springer, New York (2010)
7. Ito, T., Kanamori, R., Takahashi, J., Maestre, I.M., de la Hoz, E.: The comparison of stigmergy strategies for decentralized traffic congestion control: Preliminary results. In: In the Proceedings of The 12th Pacific Rim International Conference on Artificial Intelligence (PRICAI2012) (2012)
8. Kanamori, R., Ito, T., Takahashi, J.: Evaluation of anticipatory stigmergy strategies for traffic management. In: Proceedings of IEEE Vehicular Networking Conference (VNC) (2012)
9. Morikawa, T., Miwa, T.: Preliminary analysis on dynamic route choice behavior using probe-vehicle data. J. Adv. Transport. (2006)
10. Narzt, W., Wilflingseder, U., Pomberger, G., Kolb, D., Hortner, H.: Self-organising congestion evasion strategies using ant-based pheromones. IET Intell. Transport Syst. **4**(1), 93–102 (2010)
11. Pillac, V., Gendreau, M., Gueret, G., Medaglia, A.L.: A review of dynamic vehicle routing problems. Technical Report CIRRELT-2011-62 (2011)
12. Sheffi, Y.: Urban Transportation Networks: Equilibrium Analysis with Mathematical Programming Methods, vol. 1. Prentice-Hall, Englewood Cliffs (1985)
13. Thomas, B.W., III, C.C.W.: Anticipatory route selection. Transport. Sci. (2004)
14. Vasirani, M., Ossowski, S.: A market-based approach to accommodate user preferences in reservation-based traffic management. Technical Report ATT (2010)
15. Wolfram, S.: Theory and Application of Cellular Automata. World Scientific, Singapore (1986)

Index

A

acceptability, 91, 93, 94
agent-based simulation, 16
algorithm, 87
allocation solution, 47, 49, 52
annotations, 69
Ant Colony Optimization (ACO), 135
anticipatory stigmergy, 134, 143
anticipatory stigmergy , 144
APDL core, 68
Assistance Problem Definition Language
(APDL), 60, 65, 67, 78
assistance system (AS), 60, 62, 63, 78
auction mechanism design, 124
authority, 2, 3, 5, 7
authority propagation, 7
authority score, 7, 9

B

Bayesian Networks (BNs), 72
benchmark, 49, 52
bi-polarization global dynamic pattern, 37
Boolean-valued function, 44
Brier score, 117
broker, 84, 95

C

cellular automaton model, 138
central controller, 42
chance discovery, 99
cognitive cluster, 105
combinatorics, 47
concession, 87

consumer agent (CA), 113
context map, 102, 108
Continuous Ranked Probability Score, 130
Continuous Ranked Probability Score (CRPS),
117
continuous ranked probability score (CRPS),
117, 118, 123
coordination, 41, 56
coordinator, 43–46, 48, 49, 52, 55
cost optimization, 121

D

DARPA coordinators program, 56
data crystallization, 100
Data Set Template Processor (DSTP), 22, 23
data synthesis, 104
decay function, 6, 9
demand-side management (DSM), 114, 115
Dynamic Bayesian Networks (DBNs), 72
dynamic discovery, 101

E

e-market, 83
electricity, 85
electricity provider agent (EP), 113
Epinions trust network, 37
EvAAL competition, 61, 64
extension mechanism, 69

F

F-max-sum algorithm, 42, 49, 52
formal semantics, 71

© Springer Japan 2015
Q. Bai et al. (eds.), *Smart Modeling and Simulation for Complex Systems*, Studies
in Computational Intelligence 564, DOI 10.1007/978-4-431-55209-3